Brownian Movement
and
Molecular Reality

Brownian Movement
and
Molecular Reality

JEAN PERRIN

*(Professeur de Chimie Physique,
Faculté des Sciences,
Université de Paris)*

Translated from the
ANNALES DE CHIMIE ET DE PHYSIQUE, 8me Series,
September 1909
by
F. SODDY, M. A., F.R.S.

DOVER PUBLICATIONS, INC.
Mineola, New York

DOVER PHOENIX EDITIONS

Bibliographical Note

This Dover edition, first published in 2005, is an unabridged republication of the work originally published by Taylor and Francis, London, 1910.

International Standard Book Number: 0-486-44257-8

Manufactured in the United States of America
Dover Publications, Inc., 31 East 2nd Street, Mineola, N.Y. 11501

Brownian Movement and Molecular Reality

BROWNIAN MOVEMENT AND MOLECULAR REALITY.

By M. JEAN PERRIN

(Professeur de Chimie Physique, Faculté des Sciences, Université de Paris.)

TRANSLATED FROM THE

ANNALES DE CHIMIE ET DE PHYSIQUE, 8me SERIES,
September 1909,

BY F. SODDY, M.A., F.R.S.

I.

1. The first indication of the phenomenon.—When we consider a fluid mass in equilibrium, for example some water in a glass, all the parts of the mass appear completely motionless to us. If we put into it an object of greater density it falls and, if it is spherical, it falls exactly vertically. The fall, it is true, is the slower the smaller the object; but, so long as it is visible, it falls and always ends by reaching the bottom of the vessel. When at the bottom, as is well known, it does not tend again to rise, and this is one way of enunciating Carnot's principle (impossibility of perpetual motion of the second sort).

These familiar ideas, however, only hold good for the scale of size to which our organism is accustomed, and the simple use of the microscope suffices to impress on us new ones which substitute a kinetic for the old static conception of the fluid state.

Indeed it would be difficult to examine for long preparations in a liquid medium without observing that all the particles situated in the liquid instead of assuming a regular movement of fall or ascent, according to their density, are, on the contrary, animated with a perfectly irregular movement. They go and come, stop, start again, *mount*, descend, *remount again*, without in the least tending toward immobility. This is the *Brownian movement*, so named in memory of the naturalist Brown, who described it in 1827 (very shortly after the discovery of the achromatic objective), then proved that the movement was not due to living animalculæ, and recognised that the particles in suspension are agitated the more briskly the smaller they are.

2. Projection of the Brownian movement —This phenomenon can be made visible to a whole audience by projection, but this is difficult, and it may be useful to detail the precautions which have enabled me to arrive at a satisfactory result. The image of an electric arc (or better, of the sun) is formed in the preparation, the greater part of the non-luminous heat rays being stopped by means of a cell full of water. The rays, reflected by the particles in suspension, traverse, as for direct observation, an immersion objective and an eyepiece of high magnification, and are then turned horizontally by a total-reflection prism so as to form the image of the granules on a screen of ground glass (ruled in squares by preference, so as to have reference marks), on the farther side of which the audience is. The light is thus better utilised than with an ordinary screen which would diffuse a large part of it in directions where there were no observers. The magnification can be usefully raised to 8,000 or 10,000 diameters.

But it is necessary above all to procure an appropriate emulsion. In the few trials of projection which have been made up till now, the diameter of the granules employed was of the order of a micron, and their image is visible only with difficulty beyond 3 metres (at least with the light of the arc) whether immersion or lateral illumination is used. Smaller granules are still less visible, and one is led to this, at first sight, paradoxical conclusion, that it is better to project large granules than small ones. It is true that their movement is less, but it is still quite sufficient for its essential characteristics to be easily recognised.

It is still necessary to know how to prepare particles

having a diameter of several microns, and we shall see soon that this is equally desirable in regard to certain points in the experimental study proper of the Brownian movement. I shall indicate later (No. 32) how I have succeeded in obtaining large, perfectly spherical granules of gamboge and mastic. With such granules the Brownian movement can still be perceived at a distance of 8 or 10 metres from the screen in a hall which has been made absolutely dark.

3. Persistance of the phenomenon in absence of all causes external to the fluid. Its explanation by the movements of molecules.—The singular phenomenon discovered by Brown did not attract much attention. It remained, moreover, for a long time ignored by the majority of physicists, and it may be supposed that those who had heard of it thought it analogous to the movement of the dust particles, which can be seen dancing in a ray of sunlight, under the influence of feeble currents of air which set up small differences of pressure or temperature. When we reflect that this apparent explanation was able to satisfy even thoughtful minds, we ought the more to admire the acuteness of those physicists, who have recognised in this, supposed insignificant, phenomenon a fundamental property of matter.

Besides, as happens most frequently when it is sought to unravel the genesis of a great directing idea, it is difficult to fix precisely how the hypothesis, which ascribes the Brownian movement to molecular agitation, first appeared and how it was developed.

The first name which calls for reference in this respect is, perhaps, that of Wiener, who declared at the conclusion of his observations, that the movement could not be due to convection currents, that it was necessary to seek for the cause of it in the liquid itself, and who, finally, almost at the commencement of the development of the kinetic theory of heat, divined that molecular movements were able to give the explanation of the phenomenon [*].

Some years later Fathers Delsaulx and Carbonnelle

[*] *Erklärung des atomistischen Wesens des flüssigen Körperzustandes und Bestätigung desselben durch die sogennanten Molekularbewegungen* (*Pogg. Ann.* 1863, cxviii. 79).

published in the *Royal Microscopical Society* and in the *Revue des Questions scientifiques*, from 1877 to 1880, various Notes on the *Thermodynamical Origin of the Brownian Movement*[*]. In a note by Father Delsaulx, for example, one may read : "the agitation of small corpuscles in suspension in liquids truly constitutes a general phenomenon," that it is " henceforth natural to ascribe a phenomenon having this universality to some general property of matter," and that " in this train of ideas, the internal movements of translation which constitute the calorific state of gases, vapours and liquids, can very well account for the facts established by experiment."

In another Note, by Father Carbonnelle, one, again, may read this : " In the case of a surface having a certain area, the molecular collisions of the liquid which cause the pressure, would not produce any perturbation of the suspended particles, because these, as a whole, urge the particles equally in all directions. But if the surface is of area less than is necessary to ensure the compensation of irregularities, there is no longer any ground for considering the mean pressure ; the inequal pressures, continually varying from place to place, must be recognised, as the law of large numbers no longer leads to uniformity ; and the resultant will not now be zero but will change continually in intensity and direction. Further, the inequalities will become more and more apparent the smaller the body is supposed to be, and in consequence the oscillations will at the same time become more and more brisk . . . . . ."

These remarkable reflections unfortunately remained as little known as those of Wiener. Besides it does not appear that they were accompanied by an experimental trial sufficient to dispel the superficial explanation indicated a moment ago ; in consequence, the proposed theory did not impress itself on those who had become acquainted with it.

On the contrary, it was established by the work of M. *Gouy* (1888), not only that the hypothesis of molecular agitation gave an admissible explanation of the Brownian movement, but that no other cause of the movement could

---

[*] *See* for this bibliography an article which appeared in the *Revue des Questions scientifiques*, January 1909, where M. Thirion very properly calls attention to the ideas of these *savants*, with whom he collaborated.

be imagined, which especially increased the significance of the hypothesis *. This work immediately evoked a considerable response, and it is only from this time that the Brownian movement took a place among the important problems of general physics.

In the first place, M. Gouy observed that the Brownian movement is not due to vibrations transmitted to the liquid under examination, since it persists equally, for example, at night on a sub-soil in the country as during the day near a populous street where heavy vehicles pass. Neither is it due to the convection currents existing in fluids where thermal equilibrium has not been attained, for it does not appreciably change when plenty of time is given for equilibrium to be reached. Any comparison between Brownian movement and the agitation of dust-particles dancing in the sunlight must therefore be set aside. In addition, in the latter case, it is easy to see that the neighbouring dust-particles move in general in the same sense, roughly tracing out the form of the common current which bears them along, whereas the most striking feature of the Brownian movement is the absolute independence of the displacements of neighbouring particles, so near together that they pass by one another. Lastly, neither can the unavoidable illumination of the preparation be suspected, for M. Gouy was able abruptly to reduce it a thousand times, or to change its colour considerably, without at all modifying the phenomenon observed. All the other causes from time to time imagined have as little influence; even the nature of the particles does not appear to be of any importance, and henceforward it was difficult not to believe that these particles simply serve to reveal an internal agitation of the fluid, the better the smaller they are, much as a cork follows better than a large ship the movements of the waves of the sea.

Thus comes into evidence, in what is termed a *fluid in equilibrium*, a property eternal and profound. This equilibrium only exists as an average and for large masses; it is a statistical equilibrium. In reality the whole fluid is

* *Journal de Physique*, 1888, 2nd Series, vii. 561; *Comptes rendus*, 1889, cix. 102; *Revue générale des Sciences*, 1895, 1.

agitated indefinitely and *spontaneously* by motions the more violent and rapid the smaller the portion taken into account; the statical notion of equilibrium is completely illusory.

4. Brownian movement and Carnot's principle.—There is therefore an agitation maintained indefinitely without external cause. It is clear that this agitation is not contradictory to the principle of the conservation of energy. It is sufficient that every increase in the speed of a granule is accompanied by a cooling of the liquid in its immediate neighbourhood, and likewise every decrease of speed by a local heating, without loss or gain of energy. We perceive that thermal equilibrium itself is also simply a statistical equilibrium. But it should be noticed, and this very important idea is again due to M. Gouy, that the Brownian movement is not reconcilable with the rigid enunciations too frequently given to Carnot's principle; the particular enunciation chosen can be shown to be of no importance. For example, in water in equilibrium it is sufficient to follow with the eyes a particle denser than water to see it at certain moments rise spontaneously, absorbing, necessarily, work at the expense of the heat of the surrounding medium. So it must not any longer be said that perpetual motion of the second sort is impossible, but one must say : "On the scale of size which interests us practically, perpetual motion of the second sort is in general so insignificant that it would be absurd to take it into account." Besides such restrictions have long been laid down : the point of view that Carnot's principle expresses simply a law approximated to has been upheld by Clausius, Maxwell, Helmholtz, Boltzmann, and Gibbs, and in particular may be recalled the *demon*, imagined by Maxwell, which, being sufficiently quick to discern the molecules individually, made heat pass at will from a cold to a hot region without work. But since one is limited to the intervention of invisible molecules, it remained possible, by denying their existence, to believe in the perfect rigidity of Carnot's principle. But this would no longer be admissible, for this rigidity is now in opposition to a *palpable reality*.

On the other hand, the practical importance of Carnot's principle is not attacked, and I hardly need state at length that it would be imprudent to count upon the Brownian movement to lift the stones intended for the building of a house. But the comprehension of this important principle becomes in consequence more profound : its connection with the structure of matter is better understood, and the conception is gained that it can be enunciated by saying that spontaneous co-ordination of molecular movements becomes the more improbable the greater the number of molecules and the greater the duration of time under consideration *.

5. The kinetic molecular hypothesis.—I have said that the Brownian movement is explained, in the theory of M. Gouy and his predecessors, by the incessant movements of the molecules of the fluid, which striking unceasingly the observed particles, drive about these particles irregularly through the fluid, except in the case where these impacts exactly counterbalance one another. It has, to be sure, been long recognised, especially in explanation of the facts of diffusion, and of the transformation of motion into heat, not only that substances in spite of their homogeneous appearance, have a discontinuous structure and are composed of separate *molecules*, but also that these molecules are in incessant agitation, which increases with the temperature and only ceases at absolute zero.

Instead of taking this hypothesis ready made and seeing how it renders account of the Brownian movement, it appears preferable to me to show that, possibly, it is logically suggested by this phenomenon alone, and this is what I propose to try.

What is really strange and *new* in the Brownian movement is, precisely, that it never stops. At first that seems in contradiction to our every-day experience of friction. If for example, we pour a bucket of water into a tub, it seems natural that, after a short time, the motion possessed by the liquid mass disappears. Let us analyse further how this apparent equilibrium is arrived at : all the particles had at

---

* With regard to the general significance of the principle I should refer to the very interesting considerations developed by J. H. Rosny, Senior, in his book on Pluralism, pp. 85-91 (F. Alcan, 1909).

first velocities almost equal and parallel ; this co-ordination is disturbed as soon as certain of the particles, striking the walls of the tub, recoil in different directions with changed speeds, to be soon deviated anew by their impacts with other portions of the liquid. So that, some instants after the fall, all parts of the water will be still in motion, but it is now necessary to consider quite a small portion of it, in order that the speeds of its different points may have about the same direction and value. It is easy to see this by mixing coloured powders into a liquid, which will take on more and more irregular relative motions.

What we observe, in consequence, so long as we can distinguish anything, is not a cessation of the movements, but that they become more and more chaotic, that they distribute themselves in a fashion the more irregular the smaller the parts.

Does this de-co-ordination proceed indefinitely ?

To have information on this point and to follow this de-co-ordination as far as possible after having ceased to observe it with the naked eye, a microscope will be of assistance, and microscopic powders will be taken as indicators of the movement. Now these are precisely the conditions under which the Brownian movement is perceived: we are therefore *assured* that the de-co-ordination of motion, so evident on the ordinary scale of our observations, does not proceed indefinitely, and, on the scale of microscopic observation, we *establish* an equilibrium between the co-ordination and the de-co-ordination. If, that is to say, at each instant, certain of the indicating granules stop, there are some in other regions at the same instant, the movement of which is re-co-ordinated automatically by their being given the speed of the granules which have come to rest. So that it does not seem possible to escape the following conclusion :

Since the distribution of motion in a fluid does not progress indefinitely, and is limited by a spontaneous re-co-ordination, it follows that the fluids are themselves composed of granules or *molecules*, which can assume all possible motions relative to one another, but in the interior of which dissemination of motion is impossible. If such molecules had no existence it is not apparent how there would be any limit to the de-co-ordination of motion.

On the contrary if they exist, there would be, unceasingly, partial re-co-ordination ; by the passage of one near another, influencing it (it may be by *impact* or in any other manner), the speeds of these molecules will be continuously modified, in magnitude and direction, and from these same chances it will come about sometimes that neighbouring molecules will have concordant motions. In addition, even without this absolute concordance being necessary, it will at least come about frequently that the molecules in the region of an indicating particle will assume in a certain direction an excess of motion sufficient to drive the particle in that direction.

The Brownian movement is permanent at constant temperature : that is an experimental fact. The motion of the molecules which it leads us to imagine is thus itself also permanent. If these molecules come into collision like billiard balls, it is necessary to add that they are perfectly elastic, and this expression can, indeed, be used to indicate that in the molecular collisions of a thermally isolated system the sum of the energies of motion remains definitely constant.

In brief the examination of Brownian movement alone suffices to suggest that every fluid is formed of elastic molecules, animated by a perpetual motion.

6. **The atoms. Avogadro's constant.**—From this, as is well known, diverse considerations of chemistry, and particularly the study of substitution, lead to the idea of the existence of atoms. When, for example, calcium is dissolved in water, only one half of the hydrogen contained in the latter is displaced. The hydrogen of this water, and in consequence the hydrogen of each molecule, is therefore composed of two distinct parts. No experiments lead to any further differentiation, and it is reasonable to regard these two parts as indivisible, by all chemical methods, or in a word, they are *atoms*. On the other hand, every mass of water, and in consequence each molecule of water, weighs 9 times the hydrogen it contains : the molecule of water, which contains

2 atoms of hydrogen, weighs therefore 18 times the atom of hydrogen. In a similar manner, it may be established that the molecule of methane, for example, weighs 16 times more than the atom of hydrogen. Thus, by a purely chemical method, through the conception of the atom, the ratio 16/18, of the weight of a molecule of methane to a molecule of water, can be reached.

Now this same ratio, precisely, is arrived at by comparison of the masses of similar volumes of methane and water vapour in the gaseous state under similar conditions of temperature and pressure. Thus these two masses, which have the same ratio as the two kinds of molecules, must contain as many molecules the one as the other. This result is general for the different gases, so that in consequence we arrive, in an experimental manner, at the celebrated proposition enunciated in the form of an hypothesis by Avogadro, about a century ago, and taken up again a little later by Ampère:

"Any two gases, taken under the same conditions of temperature and pressure, contain in the same volume the same number of molecules."

It has become customary to name as the gram-molecule of a substance, the mass of the substance which in the gaseous state occupies the same volume as 2 grams of hydrogen measured at the same temperature and pressure. Avogadro's proposition is then equivalent to the following:

"*Any two gram-molecules contain the same number of molecules.*"

This invariable number N is a universal constant, which may appropriately be designated *Avogadro's Constant*. If this constant be known, the mass of any molecule is known: even the mass of any atom will be known, since we can learn by the different methods which lead to chemical formulæ, how many atoms of each sort there are in each molecule. The weight of a molecule of water, for example, is $\frac{18}{N}$; that of a molecule of oxygen is $\frac{32}{N}$, and so on for each molecule. Similarly the weight of the oxygen atom, obtained by dividing the gram-atom of oxygen by N, is $\frac{16}{N}$; that of the atom of hydrogen is $\frac{1 \cdot 008}{N}$, and so on for each atom.

7. The constant of molecular energy.—It is easy to see that if we know Avogadro's constant we can calculate the mean kinetic energy of translation of different molecules, and conversely, the value of this energy will give us N. Let us elaborate this important point a little.

If fluids are composed of molecules in motion, the pressure which they exert on the boundaries which limit their expansion is accounted for by the impacts of their molecules against these boundaries, and, in the case of gases (the molecules of which are very remote, relatively, one from another), it has been established, thanks to the successive arguments, created or modified by Joule, Clausius, and Maxwell, that this conception, at first somewhat vague, contains the precise relation

$$pv = \frac{2}{3}nw$$

where p is the pressure which n molecules of mean kinetic energy w develop in the volume v.

If the mass of gas under consideration is one gram-molecule, n becomes equal to N and pv to RT, T being the absolute temperature and R the constant of a perfect gas (equal in C.G.S. units to $83 \cdot 2 \times 10^6$); the preceding equation may then be written

$$\frac{2}{3}Nw = RT$$

or

$$w = \frac{3R}{2N}T.$$

Now the constant N is the same for all substances. The molecular kinetic energy of translation has thus for all gases the same mean value, proportional to the absolute temperature

$$w = \alpha T.$$

The constant α, which may be named the *constant of molecular energy*, equal to $\frac{3R}{2N}$, is, like N, a universal constant.

It is evident that both of these constants will be known as soon as one is.

8. The atom of electricity.—A third universal constant is also reached at the same time as N or α, and this is encountered in the study of the phenomena of electrolysis. It is known that the *decomposition* by the current of the gram-molecule of a given electrolyte is accompanied always by the passage of the same quantity of electricity: as is well known, this is explained by the conception that in all electrolytes a part at least of the molecules are dissociated into *ions* carrying fixed electric charges, and in consequence sensitive to the electric field; lastly, if the name *faraday* is given to the quantity F of electricity (96,550 coulombs) which passes in the decomposition of 1 gram-molecule of hydrochloric acid, it is known that the decomposition of any other gram-molecule is accompanied by the passage of a whole number of faradays, and, in consequence, that any ion carries a whole number of times the charge on the hydrogen ion. This charge e thus also appears as indivisible, and constitutes the atom of electricity or the electron (Helmholtz).

It is easy to obtain this universal constant if either of the constants, N or α, is known. Since the gram-atom of hydrogen in the ionic state, that is to say N atoms of hydrogen, carries one faraday, then necessarily,

$$Ne = F,$$

which is, in c.g.s. electrostatic units,

$$Ne = 96{,}550 \times 3.10^9 = 29.10^{13};$$

thus, in the same step, the three universal constants N, e, α will be found. Can this be accomplished?

9. Molecular speeds. Maxwell's law of irregularities. Mean free path.—The commencement of the answer to this question, and, at the same time, the approximate determination of the order of molecular magnitude, is due to the admirable efforts of Clausius, Maxwell, and Van der Waals. Without entering into detail I think it useful to summarise the line they have followed.

First, for each gas the mean square, U^2, of the molecular speed is easily calculated from the equation just written

$$\frac{2}{3} N w = RT.$$

It is sufficient to notice that $2Nw$ can be replaced by MU^2, M representing the gram-molecule of the gas under consideration. Thus it is found that U is of the order of some hundreds of metres per second (435 metres at $0°$ for oxygen).

As well understood, the molecular speeds are very variable and unequal; but in a steady state the proportion of molecules which have any definite speed remains fixed. On the hypothesis that the probability of a component x is independent of the values of the components y and z, Maxwell succeeded in determining the law of distribution of molecular speeds. His reasoning demonstrated that, on this hypothesis, the probability of any molecule possessing, along the axis Ox, a component between x and $x+dx$ had the value

$$\frac{1}{U} \sqrt{\frac{3}{2\pi}} e^{-\frac{3}{2}\frac{x^2}{U^2}} dx.$$

This expression represents the irregularities of molecular motion. It is obtained just the same on the hypothesis that the components along the Ox axis are distributed around the zero value according to the so-called law of *chance* enunciated by Laplace and Gauss.

This *law of the distribution of velocities* permits the calculation of the mean speed Ω, which is not equal to U (any more than $\frac{a+b}{2}$ is equal to the square root of $\frac{a^2+b^2}{2}$), but which, as a matter of fact, differs but little from it

$$\left(\Omega = U \sqrt{\frac{8}{3\pi}}\right).$$

On the other hand, this same law of distribution can be used to test by calculation the hypothesis that the *internal friction* between two parallel layers of gas, moving at different speeds, results from the continuous arrival, in each layer, of molecules coming from the other layer. Maxwell in this way found that the coefficient ζ of internal friction, or viscosity, which is experimentally measurable, should be very nearly equal to one-third of the product of the following three quantities : the absolute density δ of the gas (given by the balance), the mean molecular speed Ω (which we know how to calculate), and the mean free path L which a molecule traverses in a straight line between two successive impacts. More exactly, he found

$$\zeta = 0.31\, \delta\, \Omega\, L.$$

The value of the mean free path is thus obtained : for example, for oxygen or nitrogen at ordinary temperature and under atmospheric pressure it is approximately equal to 0·1 micron. At the low pressure of a Crookes' tube it can reach many centimetres.

10. The relation of the mean free path of the molecule to its diameter.—In addition, a line of reasoning, due to Clausius, shows that this same mean free path can be calculated in another manner as a function of the nearness of approach of the molecules and of their dimensions[*]. It is easy to understand that the smaller the molecules are the nearer their approach, and the larger they are the more they act as obstructions.

But there are certainly other considerations to be taken into account, as, for example, that a molecule in the form of a rod (as in the case, possibly, of certain molecules of the fatty series) will not obstruct in the same way as if it had the form of a sphere. In default of any knowledge of the exact shape of molecules, it has been thought that no great error is likely to result in likening them to spherical balls, having a diameter equal to the mean distance apart of the centres of two molecules on impact. This hypothesis, possibly exact in the case of monatomic molecules (mercury, argon, etc.), is certainly false for other molecules, but it is

[*] *Pogg. Ann.* 1858.

still possible that it may lead to approximate results in the case of the less complicated molecules such as those of oxygen and nitrogen.

Let us then liken the molecules to spheres. The approximate calculation of Clausius, subsequently modified by Maxwell, showed that the following relation should hold approximately,

$$L = \frac{1}{\pi\sqrt{2}} \frac{1}{nD^2},$$

where D represents the molecular diameter, and n the number of molecules contained in each cubic centimetre. Since L can be calculated, a second relation between n and D will give us the diameter of the molecules and the number n per cubic centimetre. In this case, multiplying the number n by the known volume of the gram-molecule under the conditions of temperature and pressure chosen in the calculation, we shall have the number N of molecules in a gram-molecule, *that is to say the required three universal constants.*

But this second relation between n and D has not been very easy to obtain.

11. First determinations of Avogadro's constant.—To begin with, the molecules in the liquid state cannot be more closely packed than bullets are in a pile of bullets [*]. Now it is easily established that the volume of bullets is only equal to 73 per cent. of the volume of the pile. So, we have

$$\frac{1}{6}\pi n D^3 < 0{\cdot}73\,\phi,$$

where ϕ signifies the known volume occupied by the mass of a cubic centimetre of the gas considered in the liquid state at low temperature. This inequality combined with the preceding equation, which gives the product nD^2, leads to a value certainly too great for the molecular diameter, and therefore to values certainly too small for n and N.

The calculation is usually made for oxygen (which gives

---

[*] In reality, the original reasoning, due to Loschmidt, was limited to the statement that the volume of the molecules is inferior to that of the liquid and perhaps not more than ten times smaller.

N > 9.10^{22}): it is better to make it for a monatomic gas, for which the molecules may really be spherical, and recommencing the calculation for mercury, the mean free path of which at 370° is 21.10^{-6} (Landolt's Tables), I find for the inferior limit of N a higher and therefore more useful value, namely

$$N > 45.10^{22}.$$

As for the molecular diameter, for all the gases considered it is found to be less than the millionth of a millimetre (for the special case of mercury $D < 3.5 \times 10^{-8}$).

This indication only puts us in the same position as that of an astronomer who, desiring to know the distance of a star from the Sun, finds at first only that it is farther off from it than Neptune. Failing a precise measurement, at least it is desirable to close in this star between two limits and to know for example whether it is nearer than Sirius.

This second limit can be fixed from a theory of dielectrics due to Clausius and Mossotti: on this theory the dielectric power of a gas depends upon the polarisation of each molecule by influence by the displacement of interior electric charges. Developing this hypothesis, we shall write that the true volume of n molecules is not equal (as is sometimes stated) but certainly greater than the volume u of n perfectly conducting spheres which could be put in the place of the molecules without modifying the dielectric constant K of the medium. An electrostatic calculation gives to u the value $\frac{K-1}{K+2}$; one can thus write

$$\frac{1}{6}\pi n D^3 > \frac{K-1}{K+2}.$$

The constant K, being practically equal to the square of the refractive index (Maxwell), can also be measured directly.

Applying to the case of argon and obtaining nD^2 from Clausius's equation, we obtain

$$N < 200.10^{22}.$$

As for the molecular diameter, it is found, for all the gases so considered, to be greater than a ten-millionth of a millimetre (for the special case of argon $D > 1.6 \times 10^{-8}$).

Here are, therefore, the various molecular magnitudes confined between two limits, which as regards the weight of each molecule are to one another as 45 is to 200. That we have no better estimate is mainly because we only know how to evaluate roughly the true volume of n molecules which occupy the unit volume of gas. A more delicate analysis is due to Van der Waals, who appears to have obtained as much in connection with molecular magnitudes as the calculation from the mean free path is able to yield [*]. The gas equation was obtained by supposing the molecules sufficiently separated from one another for their true volume to be small compared with that occupied by their trajectories and that each molecule suffers no sensible influence, similar to cohesion, attracting it towards the whole of the others. Van der Waals was successful in allowing for these two neglected complications and obtained the celebrated equation

$$\left(p + \frac{a}{v^2}\right)(v-b) = \mathrm{RT},$$

approximately true for the whole fluid state, in which the particular nature of the substance studied comes into evidence through the two constants a and b, of which a represents the influence of cohesion, and b exactly four times the true volume of the molecules of the given mass. Hence once b is known, the equation

$$\frac{1}{6}\pi n \mathrm{D}^3 = \frac{b}{4}$$

in conjunction with the equation of Clausius and Maxwell

$$\pi n \mathrm{D}^2 = \frac{1}{\mathrm{L}\sqrt{2}},$$

enables the unknown n and D to be calculated.

This calculation has been made for oxygen and nitrogen and a value for N nearly equal to 45.10^{22} has been obtained: this choice of substances is not a very suitable one, since it necessitates the consideration of the *molecular diameter* for molecules assuredly not spherical. By utilising for the

---

[*] *Continuité des états liquides et gazeux*, 1873.

calculation the values recently given for argon, near its critical point, I find

$$N = 62.10^{22}.$$

I should add that it is not easy to estimate the error possibly affecting this number, because of the lack of rigour of the equation of Clausius-Maxwell and of that of Van der Waals. Unquestionably an uncertainty of 30 per cent. will not be a matter for surprise.

With the determination of Van der Waals we reach the end of the first series of efforts. By methods completely different we proceed to consider similar results for which the determination can be made with greater accuracy.

12. The equipartition of energy.—We have seen that the mean molecular energy is, at the same temperature, the same for all gases. This result remains valid when the gases are mixed. It is indeed known that each gas presses upon the enclosure *as if it alone were present*, that is to say that n molecules of this gas develop in the volume v the same partial pressure as if they were alone, in such a way that $\frac{3}{2}\frac{pv}{n}$ preserves the same value. On the other hand, when we try to repeat the reasoning which led to the relation

$$pv = \frac{2}{3}nw,$$

it is found that this reasoning remains applicable. Thus w must preserve the same value. For example, the molecules of carbon dioxide and water vapour, present in the air, must have the same mean kinetic energy in spite of the difference of their natures and their masses.

This invariability of molecular energy is not confined to the gaseous state, and the beautiful work of Van't Hoff has established that it extends to the molecules of all dilute solutions. Let us imagine that a dilute solution is contained in a *semi-permeable* enclosure, which separates it from the pure solvent: we suppose this enclosure allows free passage to the molecules of the solvent, in consequence of which these molecules cannot develop any pressure, but that it stops the dissolved molecules. The impacts of these molecules against the enclosure will then develop an *osmotic pressure* P, and it is seen, if the reasoning is considered in detail, that the pressure produced by these impacts can be calculated as in the case of a gas, so that in consequence we write

$$Pv = \frac{2}{3} n W,$$

W signifying the mean kinetic energy of translation of n molecules contained in the volume v of the enclosure.

Now Van't Hoff has observed that the experiments of Pfeffer give for the osmotic pressure a value equal to the pressure which would be exerted by the same mass of dissolved substance if it alone occupied in the gaseous state the volume of the enclosure. W is thus equal to w: the molecules of a dissolved substance have the same mean energy as in the gaseous state.

I wish to make a remark on this matter which appears to me to render intuitive an important proposition which the kinetic theory of fluids establishes in a somewhat laborious manner. Van't Hoff's law tells us that a molecule of ethyl alcohol in solution in water has the same energy as one of the molecules of vapour over the solution; it would still have this energy if it were present in chloroform (that is to say if it were surrounded by chloroform molecules), or even if it were in methyl or propyl alcohol: this indifference to the nature of the molecules of the liquid in which it moves makes it almost impossible to believe that it would not still have the same energy if it were in ethyl alcohol; that is if it forms one of the molecules of pure ethyl alcohol. It therefore follows that the molecular energy is the same in a liquid as in a gas, and we can now say:

At the same temperature all the molecules of all fluids have the same mean kinetic energy, which is proportional to the absolute temperature.

But this proposition, already so general, can be still further enlarged. According to what we have just seen, the heavy molecules of sugar which move in a sugar solution have the same mean energy as the agile molecules of water. These sugar molecules contain 35 atoms; the molecules of sulphate of quinine contain more than 100 atoms, and the most complicated and heaviest molecules to which the laws of Van't Hoff (or of Raoult which are deduced from them) can be extended may be cited. The mass of the molecule appears absolutely unlimited.

Let us now consider a particle a little larger still, itself formed of several molecules, in a word a *dust*. Will it proceed to react towards the impact of the molecules encompassing it according to a new law? Will it not comport itself simply as a very large molecule, in the sense that its mean energy has still the same value as that of an isolated molecule? This cannot be averred without hesitation, but the hypothesis at least is sufficiently plausible to make it worth while to discuss its consequences.

Here we are then taken back again to the observation of the particles of an emulsion and to the study of this wonderful movement which most directly suggests the molecular hypothesis. But at the same time we are led to render the theory precise by saying, not only that each particle owes its movement to the impacts of the molecules of the liquid, but further that the energy maintained by the impacts is on the average equal to that of any one of these molecules.

The propositions, of which I have just shown the probability, could be looked upon as special cases of the famous theorem of the *equipartition of energy* which, with Maxwell's *law of distribution of velocities*, forms the central point of the mathematical theory of molecular motion. This theorem, solved in successive steps, thanks to very numerous attempts, among which may be cited those of Maxwell, Gibbs, Boltzmann, Jeans, Langevin and Einstein, leads to the statement of the mean equality, as regards each *degree of freedom*, of the kinetic energies of translation or of rotation which are assumed in the interior of a fluid consisting of any assemblage of molecules. This theorem has had great importance even beyond the matters here broached, and, for example, it has been the means of predicting, according to the number of atoms in a molecule of gas, the ratio between the specific heats of a gas (at constant pressure and constant volume respectively). But its *proof* calls for complicated calculations, and a more simple, even if less rigorous method, has appeared to me desirable. Besides the word *theorem* should not cause illusion, for hypotheses are introduced or implied in the calculations, as in almost all theories of mathematical physics.

I need hardly say that this is not a criticism, and I think, on the contrary, that the great strength of mathematical physics, in its useful application to research and invention, consists in the bringing to light, according to correct logic (conscious or intuitive) of probabilities which qualitative reasoning would not have disclosed. As is well understood, it cannot be maintained that the theories are sufficient to establish the results they indicate without submitting them to the test of experiment.

In brief, whatever the path pursued, we are led to regard the mean energy of translation of a molecule as equal to that possessed by the granules of an emulsion. So that if we find a means of calculating this granular energy in terms of measurable magnitudes, we shall have at the same time a means of proving our theory. The experiments once made, two cases in general can present themselves. Either the numbers found will be greatly different from those given by the kinetic reasoning referred to above, and in this case, above all, if these numbers change according to the granules studied, the credit of the kinetic theories will be weakened and the origin of the Brownian movement remains undiscovered; or, the numbers will be of the order of magnitude predicted, and, in this case, not only shall we have the right to regard the molecular theory of this movement as established, but further we can seek from our experiments, a means, possibly this time exact, of determining molecular magnitudes. I hope to show that experiment has pronounced decisively in this latter direction *.

---

* My results were published in the *Comptes rendus* of May 1908 and September 1909.

II.

13. The speed of a granule in suspension is not accessible to measurement.—One method of proceeding appears direct; let us suppose one has the power of measuring the mass of a microscopic particle directly; can we not hope to obtain at least an idea of its mean speed, and in consequence of its energy, by direct readings, possibly by dividing by the time of an observation the distance between the two positions it occupies at the commencement and at the end of the observation (apparent mean speed), possibly by following its trajectory in a *camera lucida* during a given time, and then dividing by this time the total length of its trajectory?

As a matter of fact this was at first essayed*, and values can be found in different papers which are always some microns per second for the mean speed of granules of the order of a micron, and which will assign to these granules a mean energy about 100,000 times less than the kinetic theory indicated for the molecule, and this would completely overthrow the theory of the equipartition of energy.

But such values are *grossly untrue*. The entanglements of the trajectory are so numerous and rapid that it is impossible to follow them, and the trajectory seen is always infinitely shorter and more simple than the real trajectory. At the same time the apparent mean speed of a granule during a given time (the quotient of the displacement by the time) varies *absurdly* in magnitude and direction without in the least tending toward a limit when the time of observation is decreased, as can be seen in a simple manner by noting the position of a granule in a *camera lucida* from minute to minute, then, for example, every five seconds, and better still by photographing them every twentieth of a second, as Victor Henri has done, by cinematographing the movement. As is well understood, the tangent at any point of the trajectory cannot be fixed even in the roughest manner. It is one of the cases where one cannot help recalling those continuous functions which do not allow of derivation, which are regarded simply as mathematical

---

* Wiener, Ramsay, Exner, Zsigmondy and myself.

curiosities, wrongly since nature can suggest them equally as well as the derived functions.

In brief a direct method is impossible. Here is the path I have followed :

14. Extension of the laws of gases to dilute emulsions.— Let us suppose that it is possible to obtain an emulsion, with the granules all identical, an emulsion which I shall call, for shortness, *uniform*. It appeared to me at first intuitively, that the granules of such an emulsion should distribute themselves as a function of the height in the same manner as the molecules of a gas under the influence of gravity. Just as the air is more dense at sea-level than on a mountain-top, so the granules of an emulsion, whatever may be their initial distribution, will attain a permanent state where the concentration will go on diminishing as a function of the height from the lower layers, and the law of rarefaction will be the same as for the air.

A closer examination confirms this conception and gives the law of rarefaction by precise reasoning, very similar to that which enabled Laplace to correlate the altitude and the barometric pressure.

Let us imagine a uniform emulsion in equilibrium, which fills a vertical cylinder of cross section s. The state of a horizontal slice contained between the levels h and $h+dh$ would not be changed if it were enclosed between two pistons, permeable to the molecules of water, but impermeable to the granules (membranes of parchment-paper or of collodion could effectively play this part). Each of these semi-permeable pistons is subjected by the impact of the granules which it stops to an osmotic pressure. If the emulsion is dilute, this pressure can be calculated by the same reasoning as for a gas or a dilute solution, in the sense that, if at level h there are n granules per unit volume, the osmotic pressure P will be equal to $2/3n$W, if W signifies the mean granular energy ; it will be $2/3\,(n+dn)$ W at the level $h+dh$. Now the slice of granules under consideration does not fall : for this it is necessary that there should be equilibrium between the difference of the osmotic pressures, which urges it upward, and the total weight of the granules, diminished by the buoyancy of the liquid, which urges them downwards. Hence, calling ϕ the volume of each granule, Δ its density, and δ that of the intergranular liquid, we see that

$$-\frac{2}{3} s W dn = ns\, dh\, \phi(\Delta-\delta)g,$$

or

$$-\frac{2}{3} W \frac{dn}{n} = \phi(\Delta-\delta)g \,.\, dh,$$

which, by an obvious integration, involves the following relation between the concentrations n_0 and n at two points for which the difference of level is h:

$$\frac{2}{3} W \log \frac{n_0}{n} = \phi(\Delta-\delta)gh,$$

a relation which may be termed *the equation of distribution* of the emulsion. It shows clearly that *the concentration of the granules of a uniform emulsion decreases in an exponential manner as a function of the height*, in the same way as the barometric pressure does as a function of the altitude*.

If it is possible to measure the magnitudes other than W which enter into this equation, one can see whether it is verified and whether the value it indicates for W is the same as that which has been approximately assigned to the molecular energy. In the event of an affirmative answer, the origin of the Brownian movement will be established, and the laws of gases, already extended by Van't Hoff to solutions, can be regarded as still valid even for emulsions with visible granules.

15. Emulsions suitable for the researches.—Previous observations do not afford any information as to the equilibrium

---

* I indicated this equation at the time of my first experiments (*Comptes rendus*, May 1908). I have since learnt that Einstein and Smoluchowski, independently, at the time of their beautiful theoretical researches of which I shall speak later, had already seen that the exponential distribution is a necessary consequence of the equipartition of energy. Beyond this it does not seem to have occurred to them that in this sense an *experimentum crucis* could be obtained, deciding for or against the molecular theory of the Brownian movement.

distribution of the granules of an emulsion. It is only known that a large number of colloidal solutions will clarify in their upper part when they are left undisturbed for several weeks or months.

I have made some trials without result upon these colloidal solutions (sulphate of arsenic, ferric hydroxide, collargol, etc.). On the other hand, after some trials, I have been enabled to carry out measurements on emulsions of gamboge, then (with the assistance of M. Dabrowski) on emulsions of mastic.

The *gamboge*, which is used for a water-colour, comes from the desiccation of the latex secreted by *Garcinia morella* (*guttier* of Indo-China). A piece of this substance rubbed with the hand under a thin film of distilled water (as *soapsuds* can be made from a piece of soap) dissolves little by little, giving a beautiful opaque emulsion of a bright yellow colour, in which the microscope shows a swarm of yellow granules of various sizes, all perfectly spherical. These yellow granules can be separated from the liquid in which they are contained by energetic centrifuging, in the same manner as the red corpuscles may be separated from blood serum. They then collect at the bottom of the vessel centrifuged as a yellow mud, above which is a cloudy liquid which is decanted away. The yellow mud diluted anew (by shaking) with distilled water gives the mother emulsion which will serve for the preparation of the uniform emulsions intended for the measurements.

Instead of so using the natural granules the gamboge may be treated with methyl alcohol which entirely dissolves the yellow material (about four-fifths of the raw material) leaving a mucilaginous residue, to the properties of which I shall perhaps have to revert. This alcoholic solution, which is quite transparent and very similar to a solution of bichromate, changes suddenly, on the addition of much water, into a yellow opaque emulsion of the same appearance as the natural emulsion, and like it, composed of spherical granules. They can be separated again by centrifuging from the weak alcoholic liquid which contains them, then diluted with pure water, which gives, as in the preceding case, a mother emulsion which consists of granules of very different sizes, but of which the diameter is usually less than 1μ when the alcoholic and aqueous solutions are mixed without precautions.

I presume that the material so precipitated by water is a definite chemical compound and not a mixture * ; but that does not concern the end here pursued, and it is enough that the granules of the mother emulsion should have the same density, in order that it may serve for the preparation, in the manner about to be described, of uniform emulsions suitable for the measurements. Incidentally it may be said that an emulsion is *pure* when the granules forming it have the same composition (and in consequence the same density).

As regards *mastic*, which is used in the preparation of varnish, it is obtained by making incisions in the bark of the *Pistacia lentiscus* (Chios Island). It does not give an emulsion by direct manipulation with water ; but on leaving

---

* A rapid physico-chemical study has given me the following results:—

The yellow material, which is soluble in alcohol and equally very soluble in sulphide of carbon and acetic acid, is an acid for it dissolves in alkalies, even when very dilute. The acidity can be titrated by determining what quantity of soda renders a given emulsion just transparent. On the other hand, it is known that methyl alcohol and sulphide of carbon, being incompletely miscible, can give two superimposed liquid phases. The dissolved yellow colouring matter distributes itself very unequally between these two phases, and, if it were a mixture, in all probability its constituents would divide themselves unequally. Now the emulsions obtained from these two phases by addition of water require practically the same quantity of soda per gram of dissolved material to clarify them; more exactly a gram-molecule of soda dissolves 537 grams of yellow material in one emulsion and 542 grams in the other. Finally, the determination of the molecular weight by the cryoscopic method in acetic acid gives the number 555. *It is thus fairly certain that the yellow constituent of gamboge is a pure substance, with a molecular weight in the neighbourhood of* 540, *which may be termed* GUTTIC ACID (*acide guttique*).

I have further observed that this guttic acid, which expels carbon dioxide from carbonates on boiling, is displaced at ordinary temperature by a current of carbon dioxide in a solution of guttate of sodium, and then reforms an emulsion composed of spherical granules. It is thus sufficient to breathe into a clear solution of guttate to cloud it. It may be that the formation of the granules in the plant is to be explained in an analogous manner.

it in contact with methyl alcohol, there is obtained, above a completely insoluble tarry residue, a solution which is probably pure, which gives on dilution with much water an emulsion as white as milk. The granules of this emulsion are spherical and of very diverse size. The substance forming the granules has, according to Johnson, a molecular weight equal to 606, corresponding to the formula $C_{40}H_{62}O_4$.

Here are, therefore, two materials yielding spherical granules (and doubtless this would be the case for all resinous emulsions); for all such the equation of distribution of granules of radius a will be

$$\frac{2}{3} W \log \frac{n_0}{n} = \frac{4}{3} \pi a^3 (\Delta - \delta) gh,$$

or, since the Neperian logarithm is equal to the ordinary logarithm multiplied by 2·303,

$$W \, 2·303 \log_{10} \frac{n_0}{n} = 2\pi a^3 (\Delta - \delta) gh.$$

I have successively measured all the quantities which enter into this equation.

16. Fractional centrifuging. Realisation of a uniform emulsion.

—It is necessary first to know how to prepare an emulsion in which all the granules have, at least approximately, the same diameter. The procedure which I have employed can be compared to the fractionation of a liquid mixture by distillation. Just as, during distillation, the parts first vaporised are relatively richer in the more volatile constituents, so during centrifuging of a pure emulsion * the parts first deposited are relatively richer in large granules, and it was thought that this might furnish a means of separating the granules according to their size. Here is the technique which seemed the most simple to me:

The vessel of the centrifuge is filled to a definite height, 10 cm. for example, with a pure emulsion; the machine is

---

* If the primary emulsion contains granules of different densities, fractionation will always separate the granules falling from the same height in the same time, which will no longer be equal.

started at a definite angular speed, for example 30 turns per second (which gives, at 15 cm. from the axis, a centrifugal force about 500 times greater than gravity) : the motive power is cut off from the machine after a definite time (60 minutes for example), and it is allowed to come to rest of itself, which should require some minutes; the vessel is then cautiously removed.

A fairly stiff mud occupies the bottom of the vessel to a well-marked level, of a height generally negligible compared with the height of the liquid ; it contains all the granules which have reached the bottom of the vessel during centrifuging, as closely packed the one on the other, as the granules filling a bag of sand.

Let a_1 indicate the radius which a granule, initially placed at the surface, ought to have to reach the bottom of the vessel just at the moment centrifuging is stopped ; all granules larger will necessarily reach the deposited mud, but the mud will contain in addition many smaller granules, which have had time to reach the bottom because at first they were present in the lower layers of the emulsion.

By means of a siphon, the supernatant liquid is cautiously decanted off, and the vessel is refilled to the original height with distilled water ; the mud is stirred up to separate all the granules, and the preceding operation is repeated with the same angular speed and for the same time of centrifuging. All the granules, of radius above a_1, again have time to reach the bottom, but such of the smaller granules as before were able to reach the bottom because they started near it, do not do so this time if they chance to be initially near the surface. Briefly, the second sediment contains like the first all the granules of radius greater than a_1, and contains far fewer smaller granules.

The supernatant emulsion, which already is paler than the first decanted fraction, is decanted, and the same operations are repeated on the sediment until the supernatant liquid at the end of each centrifuging becomes almost as clear as water. Then this sediment contains all the granules of the primary emulsion, the radius of which exceeds a_1, and does not contain any other ; all the smaller granules have been eliminated.

Let us recommence the same operations upon this latter sediment, but with a time of centrifuging a little shorter than previously. Let a_2 indicate the radius which a granule in the surface ought to have to reach the bottom of the vessel at the end of this operation. The supernatant liquid can only contain granules of radius smaller than a_2, and owing to its origin it cannot contain granules larger than a_1; so that if a_2 is near to a_1, *this liquid is a practically uniform emulsion* which it only remains to decant off.

I do not think it is necessary to explain how, in an analogous manner, from the first fractions can be obtained at will a uniform emulsion of granules, distinctly smaller, or from the sediment a uniform emulsion of granules distinctly larger.

17. Determination of the density of the granules.—The density of the granules of the uniform emulsion upon which we wish to work must be known. I have employed two methods which give concordant results. Both make use of the fact that the mass of resin present in a given sample of emulsion can be estimated with precision by simple drying on the stove. A limiting weight is very quickly reached as soon as the temperature reaches a little above 100° and does not sensibly alter when the temperature is raised to 130° or even 140° however long the treatment lasts.

But at this temperature the resin becomes a very viscous liquid, giving on cooling a transparent glass, probably of the same density as that which formed the granules of the emulsion. We are thus led to investigate the density of this glass. This is done conveniently and exactly, by putting some fragments of it into water to which is added, little by little, potassium bromide, until the fragments remain suspended in the solution, neither rising nor falling. The required density is then equal to the density of this solution which is determined without difficulty (method of Retgers, applicable even to extremely small fragments).

The second method, while perhaps more certain, has the disadvantage of requiring much of the emulsion. At a given temperature, in the neighbourhood of 20°, the masses, m of water and m' of emulsion, are measured, which fill the same specific gravity flask. The mass μ of resin contained in this mass m' of emulsion is determined by drying, which gives the mass $(m'-\mu)$ of the intergranular water. If d is the absolute density of water the volume of the flask is $\dfrac{m}{d}$, that of the intergranular water is $\dfrac{m'-\mu}{d}$, their difference

$$\left[\frac{m}{d}-\frac{m'-\mu}{d}\right]$$

is the volume of the granules, and the quotient of their mass μ by this volume gives the density sought.

As I have said, the two methods are concordant: to take an example, the density of the granules of a certain emulsion of gamboge was found equal to 1·205 by the first method and 1·207 by the second. Also, the density of the granules of an emulsion of mastic, equal to 1·063 by the first method, was equal to 1·064 by the second.

Care must be taken that such results refer to the case where the intergranular liquid is practically pure water.

Indeed I have observed that, when this liquid contains salts, the density of the granules seems to increase, which is at once explained by the phenomenon of adsorption of the salt at the surface of the granule.

Incidentally *there is here a new method of studying adsorption and of determining the thickness of the transition layer.*

Also, if the intergranular liquid contains a colloid with invisible granules, these latter can coat the large granules of resin and change their apparent density.

This is what occurs in the natural emulsions of gamboge, where there is present (No. 15), about in the proportion of one-fifth, a colourless colloid, invisible in the ultramicroscope, which one can separate from the yellow granules by centrifuging and washing, so that they give the density obtained for the granules made from alcoholic solutions. Thus badly washed natural granules are heavier than after thorough washing. This source of error, soon recognised, slightly falsified my first determinations *.

* *Comptes rendus*, May 1908.

18. Arrangement of the observations.—It is not, as may well be understood, upon a height of some centimetres or even of some millimetres that the equilibrium distribution of the emulsions I have used can be studied, but upon the small height of a preparation arranged for microscopic observation, in the manner indicated roughly in the diagram (Fig. 1).

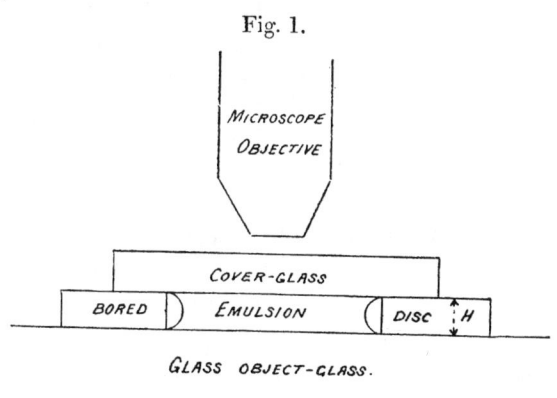

Fig. 1.

Let us suppose that a very thin glass plate bored with a large hole has been cemented in a fixed position upon a glass slide. Thus will be formed a shallow cylindrical vessel of which the height H will be, for example, about $100\,\mu$ (0·1 mm.) *.

At the centre of this vessel is placed a drop of the emulsion, which is immediately flattened by the cover-glass, and the latter, sticking to the upper face of the perforated glass plate, completely closes the cell. In addition, to prevent all evaporation, the edges of the cover-glass are covered with paraffin or varnish, which admits of a preparation being kept under observation during several days or even weeks.

The preparation is then put on to the stage of a good microscope, which has been carefully levelled. The objective used, being of very high magnifying power, has a small depth of focus, and only those granules can be seen clearly at the

---

* These requirements are quite satisfied by the *cells for enumeration of the blood corpuscles* (Zeiss), which I have employed.

same time which are present in a very thin horizontal layer, the thickness of which is of the order of a micron. By raising or lowering the microscope the granules in another layer can be seen.

The vertical distance between these two layers corresponds to the height h which enters into the equation of distribution, and this must be exactly known. We obtain it by multiplying the displacement h' of the microscope by the relative refractive index of the two media which the cover-glass separates. As the intergranular liquid is water, h will be equal to $\frac{4}{3} h'$, if a dry objective is employed, and simply equal to h' if, as I have most frequently done, a water immersion is used. As for the displacement h', it is read off directly on the graduated disc, fixed to the micrometer screw actuating the motion of the microscope (the screw of the Zeiss instrument reads to at least the quarter of a micron).

19. Counting the granules.—It is now necessary that we should be able to determine the ratio $\frac{n_0}{n}$ of the concentration of the granules at two different levels. This ratio is obviously equal to the mean ratio of the number of granules visible in the microscope at these two levels. It remains to find these numbers.

That does not at first sight appear to be easy : it is not a question of counting fixed objects, and when the eye is placed to the microscope and some hundreds of granules are seen moving in every direction, besides disappearing unceasingly while at the same time new granules make their appearance, one is soon convinced of the uselessness of attempts to estimate even roughly the mean number of granules present in the layer under observation.

The simplest course appears to be to take instantaneous photographs of this layer, to obtain the number of sharp images of granules there, and, if the emulsion is so dilute that the number is small, to repeat the process until the mean number of granules obtained on the plate can be considered known to the desired degree of approximation, for example, 1 per cent. I have, indeed, employed this proceedure for the relatively large granules, as will appear later. For granules of diameter less than 0·5 μ I have not been able to obtain good images, and I have had recourse to the following device : I placed in the focal plane of the eyepiece an opaque screen of foil pierced with a very small round hole by means of a dissecting-needle. The field of vision is thus very much diminished, and the eye can take in at a glance the exact number of granules visible at a definite instant, determined by a short signal, or during the very short period of illumination which can be obtained by means of a photographic shutter. It is necessary for this that the number does not exceed 5 or 6.

Operating thus at regular intervals, every 15 seconds for example, a series of numbers is noted down of which the mean value approaches more and more nearly a limit which gives the mean frequency of granules at the level studied, in the small cylindrical layer upon which the microscope is set. Recommencing at another level, the mean frequency is there redetermined for the same volume, and the quotient of these two numbers gives the ratio of the concentrations sought. As well understood, instead of making all the readings relating to one level continuously it is better to alternate the readings, making for example 100 at one level, then 100 at another level, then again 100 at the first level, and so on.

Some thousands of readings are required if some degree of accuracy is aimed at. To take an example, I have copied below the numbers given by 50 consecutive readings at two levels 30 μ apart in one of the emulsions I have used :—

3	2	0	3	2	2	5	3	1	2
3	1	1	0	3	3	4	3	4	4
0	3	1	3	1	4	2	2	1	3
1	1	2	2	3	0	1	3	4	3
0	2	2	1	0	2	1	3	2	4

for the lower level, and

2	1	0	0	1	1	3	1	0	0
0	2	0	0	0	0	1	2	2	0
2	1	3	3	1	0	0	0	3	0
1	0	2	1	0	0	1	0	1	0
1	1	0	2	4	1	0	1	0	1

for the upper level.

20. Determination of the radius of the granules.—To be in a position to apply the equation of distribution, we only need a single measurement, namely, that of the radius of the granules of the uniform emulsion studied. I have obtained this radius in three different ways :—

First method : At first, following the example of Sir J. J. Thomson, Langevin, and all those who in recent years have had occasion to determine the dimensions of droplets or dust-particles present in a gas, I have assumed the accuracy of the calculation of Stokes having reference to the movement of a sphere in a viscous medium. According to this calculation, the force of friction which resists the movement of a sphere is at each instant measured by $6\pi\zeta av$, if ζ indicated the viscosity of the liquid, a the radius of the sphere and v its velocity. When the sphere falls with uniform velocity under the sole influence of gravity, the force of friction must be equal to the apparent weight of the sphere in the fluid :

$$6\pi\zeta av = \frac{4}{3}\pi a^3(\Delta-\delta)g$$

an equation which gives a once the speed of fall is measured.

Let us suppose, on the other hand, that one has an extremely high vertical column of the uniform emulsion studied. It will be so far removed from the equilibrium distribution that the granules of the upper layers will fall like the droplets of a cloud, without our having practically to take into consideration the recoil due to the accumulation of granules in the lower layers. The liquid will then become clear in its upper part, and the thickness of the clear zone divided by the time during which the emulsion has been left to itself, will give the speed of fall to which the law of Stokes applies.

This phenomenon takes place in fact in the emulsions I have studied. It is sufficient to fill a *capillary* tube with the emulsion to a height of some centimetres, to seal it at its two ends and to install it vertically in a thermostat, to observe the emulsion gradually leave the upper layers of the liquip, falling like a cloud with a fairly sharp surface, and descending each day by the same amount. Fig. 2 shows the appearance observed. It is necessary to employ a capillary tube to avoid convective movements which confuse the surface of the cloud and which are produced with extreme facility in large tubes.

Fig. 2.

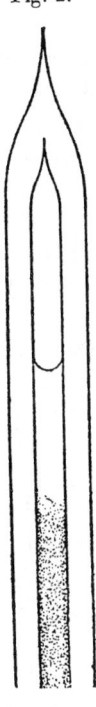

The determination of the radius of the granules is thus possible by an application of the law of Stokes. But this application to such small spheres, although definitely legitimate, gives rise to some objections which I will examine in a moment. It was thus desirable to arrive at the radius of the granules by a different, and, if possible, more direct method.

Second method.—This radius would be obtained in a very certain manner, if it were possible to find how many granules (immediately after shaking) there were in a known *titrated* volume of emulsion. That would give the mass of a granule and in consequence, since its density is known, the radius. It would be sufficient for this to count all the granules present in a cylinder of the emulsion having, as height, the height of the preparation (about 100 μ) and, as base, a surface of known area, engraved previously on the microscope slide, which is done in the *cells for the enumeration of corpuscles*, the bottom of which is divided into squares of 50 μ side. But the counting (or integration), layer by layer, of all the granules present in the height of the preparation carries with it much uncertainty. It is necessary in fact to know exactly the depth of each layer, which is of the order of a micron *, not to speak of other difficulties.

Happily I have had occasion in another connection to notice that in a feebly acid medium (for example 0·01 gram-molecule per litre) the granules of gamboge or of mastic collect on the walls of the glass which holds the preparation. At a perceptible distance from the walls the Brownian movement is in no way modified; but as soon as the chances of the movement bring a granule into contact with the slide or cover-glass, the granule becomes motionless and does not leave the wall. The emulsion is thus progressively impoverished and, after some hours, all the granules it contained are affixed to the walls. Only those, however, can be counted which are fixed in distinct positions and which do not form part of a clotted mass (partial coagulation of the colloid). Without being able to insist upon it here, I am content to say that very minute quantities of a *protecting* colloid, precisely such as is present in the natural latex of gamboge, added to the emulsion studied, prevent the granules from caking together in water acidulated by pure hydrochloric acid. On this account one may operate as follows:—

The uniform emulsion under observation, which has been previously titrated, is shaken, and a known volume of it is mixed with a known volume of feebly acidulated water, and again shaken: a drop of the mixture is taken and arranged on the microscope slide, and at once flattened by a cover-glass, the edges of which are then paraffined, taking care not to displace it, for all parts at first moistened and then

---

* We do not need to know this thickness when, in order to obtain the ratio of the concentrations at two different levels, we take the ratio of the number of granules visible at these two levels: it is sufficient for our purpose that the depth of the field, whatever it may be, has the same value for these two levels.

abandoned by the liquid carry away the granules. This done, the preparation is left on the stage of the microscope until all the granules have become attached to the walls. A *camera lucida* is then fitted to the microscope and, focussing on the bottom of the preparation, the contour which corresponds to one of the squares engraved upon the slide is drawn : the image of each of the granules fixed inside this square is marked by a point : then, adjusting the microscope until the granules fixed to the upper face are sharply defined, the images of these granules within the same contour are marked in the same way, which correspond in consequence to the same right prism of emulsion. The points on the drawing obtained can be subsequently counted at leisure, and their number is equal to the number of granules sought.

The same work is then recommenced upon another portion of the preparation, and so on until the mean value of the number of granules marked in each square can be considered well known. An obvious calculation then permits the number of granules contained in unit volume of the primary titrated emulsion to be found and gives in consequence the required radius, by a *second method* into which the law of

Fig. 3.

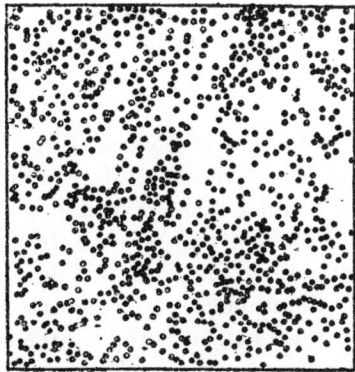

Stokes does not enter. Fig. 3 is a photograph of one of the drawings I have made on an emulsion, the granules of which had a radius equal to $0·212\,\mu$.

The use of the *camera lucida,* fatiguing in other respects, would have been avoided by directly photographing the granules fixed to the walls. But the eye is more sensitive than the photographic plate as regards the visibility of very small clear granules on a bottom almost equally clear (it must not be forgotten that the granules are transparent spheres), and I have only been able to employ photography for granules having a diameter exceeding a demimicron.

Third method.—In this case, without the granules aggregating together in irregular clots under the influence of the acid, it often happens that they arrange themselves in little rectilinear rods composed of 3, 4 or 5 granules, which can be seen moving for an instant before fixing themselves to the bottom. The length of these little rods can be measured easily in a *camera lucida* or on a photographic plate, when the diameter of a single granule (owing to magnification due to diffraction) can only be estimated in a very rough manner. This gives a *third method*, not very exact, but very direct, for the measurement of the diameter of the granules of a uniform emulsion.

For larger grains again, of the order of a micron, the regularity of the deposit becomes greater, and the granules arrange themselves one beside the other (but not one upon the other). Fig. 4 (Pl. I.) gives the photograph of the natural granules of gamboge obtained from a very nearly uniform emulsion: the direct measurement of the radius is then very easy ($0{\cdot}50\,\mu$ in the case figured).

21. Extension of the law of Stokes.—The three preceding methods give concordant results. Here is, in hundreths of a micron, the radii which they have indicated for different uniform emulsions. The numbers referring to the same emulsion occupy the same line: the middle column gives the number found by the application of the law of Stokes: the left-hand column the number found by the counting of the granules, of which the total has a known mass; the right-hand column gives the numbers found by direct measurement of the length of the columns formed by granules in juxtaposition.

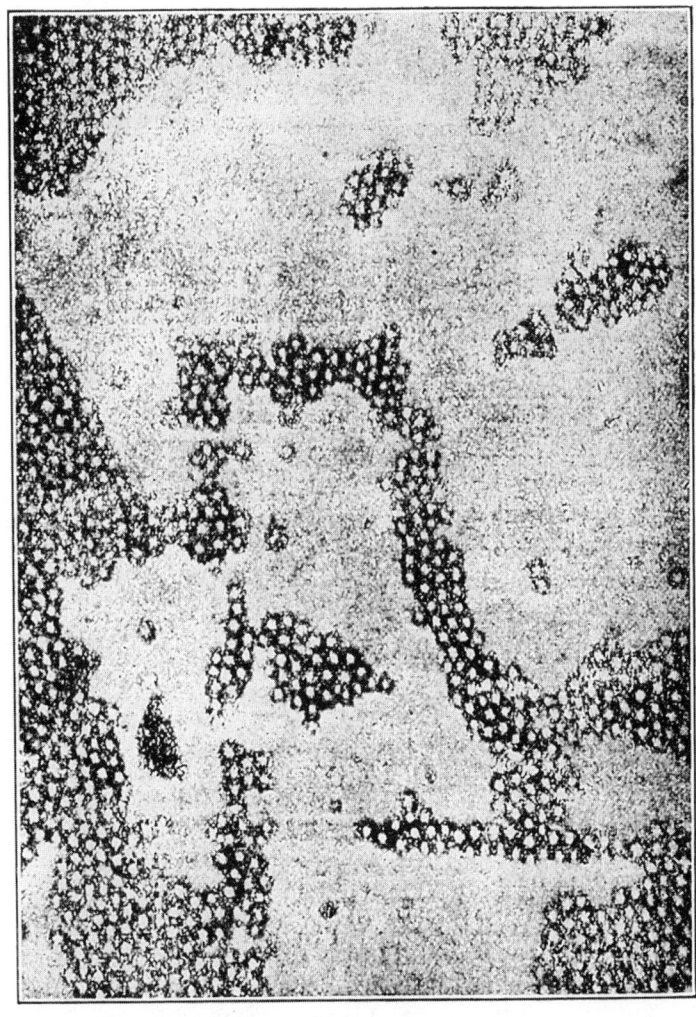

Fig. 4.

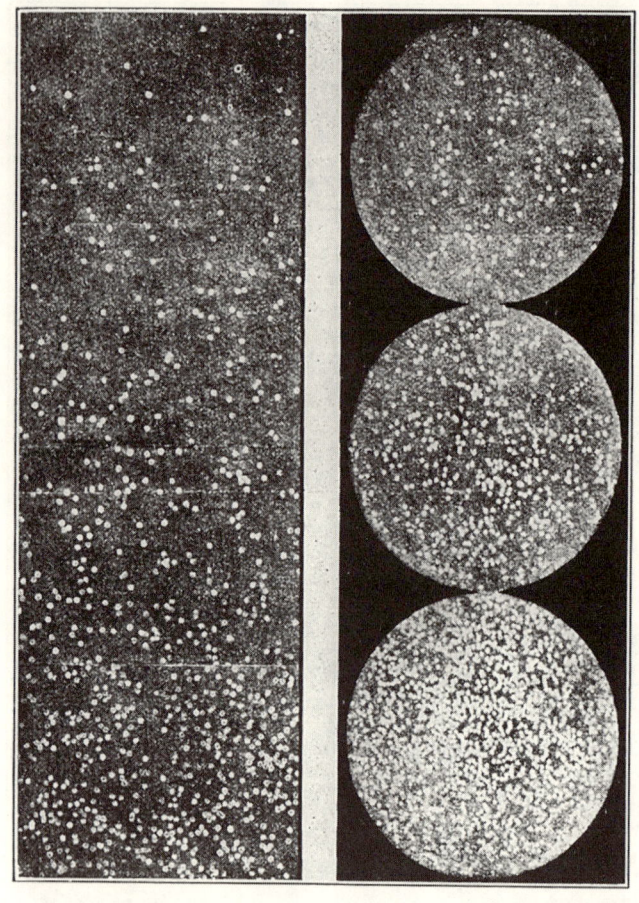

Fig. 5. Fig. 5 a.

Fig. 5. Equilibrium distribution of granules of gamboge (0·6 μ diameter; 4 levels taken every 10 μ).

Fig. 5 a. Equilibrium distribution of an emulsion of mastic (1 μ diameter; 3 levels taken every 12 μ).

	Numbering.	Law of Stokes.	Columns of granules.
Mastic.........	...	52	54
Gamboge	...	49	50
	46	45	45·5
	30	29	30
	21·2	21·3	
	14	15	

The numbers of the fifth line, which have reference to an emulsion more uniform than the others, are those which have been determined with the greatest accuracy. I have counted about 11,000 granules in different regions of different preparations to obtain the figure 21·2 of the first column.

In brief, the three methods employed justify themselves by their concordance. Further, this concordance brings out certain important consequences concerning the law of Stokes.

This law has been established by a calculation postulating conditions of continuity apparently very far from being fulfilled in the case of spheres animated by a very lively Brownian movement. Further, the speed considered is the true speed of the sphere with reference to the fluid. Now this true speed, which, as is always the case for the Brownian movement, changes unceasingly its direction and magnitude, has nothing in common with the constant vertical velocity, so incomparably smaller, with which the cloud, formed of a great number of granules, falls through the liquid. So long as I had determined the radius of these granules only by the formula of Stokes, it was still perfectly legitimate to make reservations on the accuracy of the results obtained by a risky method, as Jacques Duclaux observed in connection with my first publication [*]. Naturally the same reservations should be extended to all the cases where the formula of Stokes has been applied in the same manner; in particular to the celebrated results which the school of Sir J. J. Thomson have obtained for the condensation of droplets of water on the ions, droplets of the order of a micron, and *even more strongly* to the researches of Langevin, or his followers, on

---

[*] J. PERRIN, *Comptes rendus*, 1908, cxlvi. 967, and J. DUCLAUX, *ibid.* cxlvii. 131.

the *large ions* of the atmosphere, of which the size of the particle is of the order of the hundredth of a micron.

The concordance of the preceding measurements will dispel these doubts; but, precisely because it is not *a priori* evident, it contains something new and gives us the right to regard as experimentally established the following proposition, which is an extrapolation from the law of Stokes :

When a force, constant in magnitude and direction, acts in a fluid on a granule agitated by the Brownian movement, the displacement of the granule, which is perfectly irregular at right angles to the force, takes in the direction of the force a component progressively increasing with the time and in the mean equal to $\frac{Ft}{6\pi\zeta a}$, F *indicating the force, t the time, ζ the viscosity of the fluid, and a the radius of the granule.*

The preceding experiments show that this law is valid in the domain of microscopic quantities, and the verification, pushed even to the threshold of ultramicroscopic magnitudes, scarcely leaves a doubt that the law may be still valid for the far smaller granules of ordinary colloids, or for the *large ions* found in gases.

I presume that it still holds for molecules as large as sulphate of quinine, but I doubt if this extension can remain rigorous for molecules of radius smaller than, or but little larger than, that of the molecules of the solvent. Later on, a reason of an experimental kind will be advanced (No. **36**); but it may be now observed that the formula indicates a zero friction for a zero radius, while the real friction, which depends upon the probability of encounters between the granule under consideration and the molecules of the solvent, can only disappear if the latter also at the same time become infinitely small.

Further than this the extreme limit to which the law of Stokes can be extended does not concern the end here pursued, and, being now in possession of all the means of measurement which are necessary for the verification and utilisation of the equation of distribution of a uniform emulsion, we pass on to see what results these means give and at the same time to settle the question of the origin of the Brownian movement.

22. The progressive rarefaction as a function of the height.—
Let us consider a vertical cylinder of the emulsion, arranged
for microscopic observation in the manner detailed in No. 18.
At first, after the shaking which necessarily accompanies the
manipulation, the granules of this emulsion have an almost
uniform distribution. But, if our kinetic theory is exact,
this distribution will change from the time the preparation is
left at rest, will attain a limiting state, and in this state the
concentration will decrease in an exponential manner as a
function of the height.

This is just what experiment verifies. At first practically
as many granules are visible in the upper layers as in the
lower layers of the emulsion. A few minutes suffice for the
lower layers to become manifestly richer in granules than
the upper layers. If then the counting of the granules
(No. 19) at two different levels is commenced, the ratio $\frac{n_0}{n}$
of the concentrations at these levels is found to have a value
gradually increasing for some time, but more and more
slowly, and which ends by showing no systematic variation.
With the emulsions I have employed three hours is sufficient
for the attainment of a well-defined limiting distribution in
an emulsion left at rest, for practically the same values are
found after 3 hours as after 15 days. Those emulsions which
have not been rendered aseptic are occasionally invaded by
elongated and very active protozoa, which, by stirring up the
emulsion like fishes agitating the mud of a pond, much
diminish the inequality of distribution between the upper and
lower layers. But if one has patience to wait until these
microbes, through lack of food, die and fall inert to the
bottom of the preparation, which takes two or three days, it
will be found that the initial limiting redistribution is exactly
regained, and this possesses all the characters of the distribution of a permanent regime.

Once this permanent state is attained, it is easy to see
whether the concentration decreases in an exponential manner
as a function of the height. The following measurements
show that it is so.

At first I worked on granules of gamboge of radius
approximately equal to $0.14\,\mu$, which were studied in a cell

having a height of 110 μ. The concentrations of the granules were determined in five equidistant planes, the lowest plane being taken 5 μ above the bottom of the preparation (to eliminate the possible influence of the boundary), the distance between two consecutive planes being 25 μ, so that the uppermost plane was 5 μ below the surface.

The numbers found were between themselves as

$$100, \quad 116, \quad 146, \quad 170, \quad 200,$$

whereas the numbers

$$100, \quad 119, \quad 142, \quad 169, \quad 201,$$

which do not differ from the preceding by more than the limits of experimental error, are in geometrical progression. The distribution of the granules is thus quite exponential, as is the case for a gas in equilibrium under the influence of gravity. Only the diminution of the concentration to one-half, which for the atmosphere is produced by a height of about 6 kilometres, is produced here in a height of 0·1 millimetre.

But this fall of concentration is still too feeble for the exponential character of the decrease to be quite manifest. I have therefore tried to secure with larger granules a more rapid fall of concentration.

My most careful series has been done with granules of gamboge having a radius of 0·212 μ. The readings have been made, in a cell having a height of 100 μ, in four equidistant horizontal planes cutting the vessel at the levels

$$5 \mu, \quad 35 \mu, \quad 65 \mu, \quad 95 \mu.$$

These readings, made by direct counting through a needle-hole (No. **19**), relate to 13,000 granules and give, respectively for these levels, concentrations proportional to the numbers

$$100, \quad 47, \quad 22·6, \quad 12,$$

which are practically equal to the numbers

$$100, \quad 48, \quad 23, \quad 11·1,$$

which again are exactly in a geometrical progression.

Thus the exponential distribution cannot be doubted, each elevation of 30 μ here decreasing the concentration to about half its value.

A third series, considerably less exact, which refers to 3000 granules of an impure gamboge, more dense than the pure gamboge, has been carried out by a different method (counting of the granules on a photographic plate). The radius of the granules differed little from $0.29\,\mu$. This time an elevation of $30\,\mu$ was sufficient to lower the concentration to a tenth of its value : more precisely, the concentrations at the levels

$$5\,\mu,\quad 15\,\mu,\quad 25\,\mu,\quad 35\,\mu,$$

were between themselves as the numbers

100, 43, 22, 10,

but little different from the numbers

100, 45, 21, 9·4,

which are in geometrical progression.

In fig. 5 (Pl. II.) are shown, one above the other, drawings reproducing the distribution of the granules in four of the photographs from which the preceding numbers were obtained: the progressive rarefaction is evident. This rarefaction is very striking when, keeping the eyes fixed upon the preparation, the microscope is rapidly raised by means of its micrometer screw. The granules are then seen rapidly to become scarcer, just as the atmosphere becomes rarer around an ascending balloon, with this difference, that $10\,\mu$ in the emulsion is equivalent to 6 kilometres in the air.

I have studied emulsions of gamboge with still heavier granules. For one of these, in the lower layers, to which the measurements are limited, the concentration fell to one-fourth of its value for an elevation of $6\,\mu$. At $60\,\mu$ height, it will therefore be two million times less than at the bottom. Thus for emulsions of this kind once the permanent regime is attained, no granules are ever visible in the upper layers of the vessel employed, of height about $100\,\mu$.

Lastly, with the aid of M. Dabrowski I have established the exponential distribution in the case of emulsions of *mastic*. For example, for an emulsion where the granules had a diameter of about $1\,\mu$ ($a = 0.52\,\mu$), four photographs taken at intervals of $6\,\mu$, the one from the other, showed respectively

1880, 940, 530 and 305

images of granules, the numbers being but little different from
1880, 995, 528 and 280,
which decrease in geometrical progression. Fig. 5 *a* gives the distribution of the granules for this emulsion of mastic, in three horizontal layers placed 12 μ one above the other. There again the exponential diminution is manifest.

23. Molecular agitation is indeed the cause of the Brownian movement *.—Since the ratio of the concentrations at two points depends only on the vertical distance between them, the equation of distribution

$$2{\cdot}303 \text{ W} \log_{10}\frac{n_0}{n} = 2\pi a^3(\Delta-\delta)gh,$$

established in Nos. **14** and **15**, gives for each emulsion a well defined value of the granular energy W. If our kinetic theory is perfectly exact, this value will not depend upon the emulsion chosen, and will be equal to the mean energy w of any molecule at the same temperature. Or, what comes to the same thing, the value N' of the expression $\frac{3}{2}\frac{\text{RT}}{\text{W}}$ will be independent of the radius and density of the granules studied and will be equal to the expression $\frac{3}{2}\frac{\text{RT}}{w}$, that is (No. 7) to Avogadro's constant N, which we already know in an approximate manner (No. **11**). This is all, therefore, tantamount to seeing whether, with different emulsions, N' is placed in the neighbourhood of 60.10^{22}, the number indicated by Van der Waals' equation. The series of measurements which have enabled the exponential law of rarefaction to be established, together with others not yet described, give the answer to this question. The first series, already dealt with, refers to granules of gamboge of radius approximately equal to 0·14 μ, only moderately purified and washed, as I discovered too late, without being able since to improve the measurements as no specimen of this emulsion has been preserved. The observations, which were too difficult or impossible with

* JEAN PERRIN, *Comptes rendus*, 1908, cxlvi. 167, and cxlvii. 530.

immersion illumination, were made with lateral illumination which is appropriate for ultramicroscopic magnitudes, and referred to about 3000 granules (direct enumeration through a needle-hole). They led, allowing for the margin through the various sources of error, suitable for this series, to a value of N' lying between

$$50.10^{22} \text{ and } 80.10^{22}.$$

A second series, made with well-washed granules

$$(\Delta - \delta = 0.21),$$

of radius about double (equal approximately to $0.30\,\mu$) but only moderately uniform, also referred to about 3000 granules and gave for N' the value

$$75.10^{22}.$$

A third series, of comparable accuracy, has been done on granules about as large as the preceding ($a = 0.29\,\mu$) of an impure gamboge ($\Delta - \delta = 0.30$), by a different process (dotting the granules upon a photographic plate); it gave

$$65.10^{22}.$$

A fourth series refers to granules considerably larger ($a = 0.45\,\mu$), nearly 30 times heavier than the first granules, which were well purified ($\Delta - \delta = 0.21$). These granules were so heavy that their concentration fell to one-quarter of its value for a height of $6\,\mu$. Dotting of about 4000 granules (direct counting) gave for N'

$$72.10^{22}.$$

Lastly, it was desirable to work with a material other than gamboge. A fifth series has been done, with the aid of M. Dabrowski, on the granules of mastic. For equal volumes, these granules have an apparent weight in water three times less than the granules of pure gamboge

$$(\Delta - \delta = 0.063).$$

We have worked on granules which have a diameter a little greater than 1 micron ($a = 0.52\,\mu$) and are easily photographed.

Two plates have been taken, the one during ascent, the other during descent, at 6 equidistant horizontal layers, at intervals of $6\,\mu$. The twelve plates so obtained, on which a

total of about 7500 granules are visible, gave for N' the value
$$70.10^{22}.$$

Thus the values of the granular energy deduced from the preceding series are concordant within the limits of accuracy of the experiments, although the mass of the granules has varied as 1 : 40, the difference of density between the granules and the medium has varied as 1 : 4·7, and the rapidity of rarefaction as a function of the height has varied as 1 : 30. Incidentally, there is here, in a province which will soon be enlarged, *an experimental verification of the equal distribution of energy between different masses.*

But, further, it is manifest that these values agree with that which we have foreseen for the molecular energy. The mean departure does not exceed 15 per cent., and the number given by the equation of Van der Waals does not allow of this degree of accuracy.

I do not think this agreement can leave any doubt as to the origin of the Brownian movement. To understand how striking this result is, it is necessary to reflect that, before this experiment, no one would have dared to assert that the fall of concentration would not be negligible in the minute height of some microns, or that, on the contrary, no one would have dared to assert that all the granules would not finally arrive at the immediate vicinity of the bottom of the vessel. The first hypothesis would lead to the value *zero* for N', while the second would lead to the value *infinity*. That, in the immense interval which *a priori* seems possible for N', the number found should fall precisely on a value so near to the value predicted, certainly cannot be considered as the result of chance.

Thus the molecular theory of the Brownian movement can be regarded as experimentally established, and, at the same time, *it becomes very difficult to deny the objective reality of molecules.* At the same time we see the law of gases, already extended by Van't Hoff to dilute solutions, extended to uniform emulsions. The Brownian movement offers us, on a different scale, the faithful picture of the movements possessed, for example, by the molecules of oxygen dissolved in the water of a lake, which, encountering one another only rarely, change their direction and speed by virtue of their impacts with the molecules of the solvent.

It may be interesting to observe that the largest of the granules, for which I have found the laws of perfect gases followed, are already visible in sunlight under a strong lens. They behave as the molecules of a perfect gas, of which the gram-molecule would weigh *200,000 tons*.

I add lastly, that all the measurements detailed in this paragraph have been made on dilute emulsions, which in the parts richest in granules only contain a thousandth part of resin, and where the osmotic pressure does not reach a thousand millionth of an atmosphere.

This last figure shows to what a point I was removed from the conditions under which it has been possible to reveal (Malfitano), and then to measure (J. Duclaux) the osmotic pressure of colloidal solutions with very fine granules closely crowded together. It may be that a generalisation, more or less analogous to that of Van der Waals, will give, one day, by means of a reasoning from the kinetic theory, the osmotic pressure of such solutions.

24. Precise determination of Avogadro's constant.—Recapitulating, equal granules distribute themselves in a dilute emulsion as heavy molecules obeying the laws of perfect gases, and the equation of their distribution, since W may now be replaced by $\frac{3}{2}\frac{RT}{N}$, can be written

$$2\cdot303\frac{RT}{N}\log_{10}\frac{n_0}{n}=\frac{4}{3}\pi a^3 g(\Delta-\delta)h.$$

Once this has been well established, this same equation affords a means for determining the constant N, and the constants depending upon it, which is, it appears, *capable of an unlimited precision*. The preparation of a uniform emulsion and the determination of the magnitudes other than N which enter into the equation can in reality be pushed to whatever degree of perfection desired. It is simply a question of patience and time; nothing limits *a priori* the accuracy of the results, and the mass of the atom can be obtained, if desired, with the same precision as the mass of the Earth. I scarcely need observe, on the other hand, that even perfect measurements of compressibility might not be able to prevent an uncertainty of perhaps 40 per cent. in the value of N, deduced from the equation of Van der Waals, by means of hypotheses which we know are certainly not completely exact *.

The values found for N by the five series of experiments detailed give a rough mean of $69 \cdot 10^{22}$; the most careful of the series is the one made with *mastic* (dotting upon photographic plates) which gives $70 \cdot 10^{22}$.

I have made, with *gamboge*, a sixth series already mentioned above on various occasions, which I consider considerably more accurate still. The mean radius of the granules of the emulsion employed was found equal to $0.212\ \mu$, by counting 11,000 granules of a titriated emulsion, and to $0.213\ \mu$ by application of the law of Stokes. The difference of density between the material of the granules and the intergranular water was 0.2067 at $20°$, the temperature to which the measurements refer. 13,000 granules were counted at different heights (direct observation through a needle-hole), and it was verified that the distribution was quite exponential, each elevation of $30\ \mu$ lowering the concentration to about half of its value (exact figures are given in No. 22). The value resulting from these measurements is 70.5×10^{22}.

25. Numerical values of Avogadro's constant and of the constants depending upon it.—Thus, then, one is led to adopt for Avogadro's constant the value

$$N = 70.5 \times 10^{22}.$$

The number n of the molecules per cubic centimetre of gas in normal conditions of temperature and pressure,

---

* Sphericity of the molecules, and various simplifications in the reasoning which lead to the expression for the mean free path, make it impossible to specify exactly what uncertainty exists in the numerical coefficients of the approximate equations which connect the viscosity, the mean free path, and the molecular diameter (Nos. 9 and 10).

obtained by dividing the preceding by the volume, 22,400, of the gram-molecule, is thus

$$n = 3 \cdot 15 \times 10^{19}.$$

The constant of molecular energy (No. 6), equal to $\frac{3R}{2N}$, is, consequently, in C.G.S. units

$$\alpha = 1 \cdot 77 \times 10^{-16},$$

and in consequence the mean kinetic energy of any molecule at 0°, equal to 273 α, is, in ergs,

$$w = 0 \cdot 48 \times 10^{-13}.$$

Lastly, the charge of the electron, or atom of electricity, is obtained by dividing the faraday by N (No. 8), and is therefore in C.G.S. electrostatic units,

$$e = 4 \cdot 11 \times 10^{-10}.$$

26. Weight and dimensions of molecules or atoms.—The mass of any molecule or atom is obtained in an obvious manner with the same precision. For example, since there are in 32 grams of oxygen N molecules, each molecule of oxygen will have the mass

$$O_2 = \frac{32}{N} = 45 \cdot 4 \times 10^{-24},$$

and the atom of oxygen will be

$$O = \frac{16}{N} = 22 \cdot 7 \times 10^{-24}.$$

In the same way each molecule will be obtained by dividing by N the gram-molecule of the corresponding compound, and each atom by dividing by N the gram-atom of the corresponding element. The lightest of all the atoms, that is to say, the atom of hydrogen, has therefore the mass

$$H = \frac{1 \cdot 008}{N} = 1 \cdot 43 \times 10^{-24}.$$

Lastly, the mass of one of the identical *corpuscles* which carry the negative electricity of the cathode-rays or of the β-rays is itself obtained accurately, since it is known that it is 1775 times smaller than that of the atom of hydrogen (Classen). This corpuscular mass, the latest element of matter revealed to man, is thus

$$c = 0\cdot 805 \times 10^{-27}.$$

As for the *dimensions* of the molecules, we can, now that we know n, deduce them from the equation (No. **10**) of Clausius-Maxwell,

$$L = \frac{1}{\pi \sqrt{2}} \frac{1}{nD^2},$$

for all gases for which the mean free path L (that is to say, definitively, the viscosity) is known.

For example, at 370°, the mean free path of the molecule of mercury under atmospheric pressure is deduced from the viscosity 6.10^{-4} of the gas by Maxwell's equation

$$\eta = 0\cdot 31 \rho \Omega L,$$

which gives for L the value $2\cdot 1 \times 10^{-5}$. On the other hand, at 370°, n is equal to $3\cdot 15 \times 10^{19} \frac{273}{273+370}$. The diameter sought is therefore equal to the square root of

$$\frac{1}{\pi \sqrt{2}} \cdot \frac{1}{3\cdot 15 \times 10^{19}} \cdot \frac{643}{273} \cdot \frac{10^5}{2\cdot 1},$$

that is to say, practically $2\cdot 8 \times 10^{-8}$.

It is in this way I have calculated some of the molecular diameters, as follows :—

Helium	$1\cdot 7 \times 10^{-8}$.
Argon	$2\cdot 7 \times 10^{-8}$.
Mercury	$2\cdot 8 \times 10^{-8}$.
Hydrogen	2×10^{-8}.
Oxygen	$2\cdot 6 \times 10^{-8}$.
Nitrogen	$2\cdot 7 \times 10^{-8}$.
Chlorine	4×10^{-8}.
Ether	6×10^{-8}.

It is clear that, in the case of polyatomic molecules, one can only deal with a badly defined diameter, of which the determination, although but little affected by variations of the mass, cannot from its nature have the certainty possible for the masses.

Incidentally it may be observed that a molecule of hydrogen is insignificant compared to our own body to the same degree as that in turn is insignificant compared to the Sun.

Lastly, even the diameter of the corpuscle can be arrived at, if it is supposed, with Sir J. J. Thomson, that all its inertia is of electromagnetic origin, in which case its diameter is given by the equation *

$$D = \frac{4}{3} \frac{e^2}{mV^2},$$

where V signifies the velocity of light, m the mass of the corpuscle and e its charge, that is to say $4 \cdot 1 \times 10^{-10}$. From this there results for D the value $0 \cdot 33 \times 10^{-12}$, a value enormously smaller than the diameter of the smallest atoms.

III.

27. The formulæ of Einstein.—The preceding experiments allow, as we have seen, the origin of the Brownian movement to be established and the various molecular magnitudes to be determined. But another experimental advance was possible, and has been suggested by Einstein † at the conclusion of the very beautiful theoretical investigations of which I must now speak ‡.

We have not, so far, given a precise characterisation to the activity of the Brownian movement which agitates a definite granule, and we have only observed that its true speed is not directly measurable. Without further troubling about the infinitely tangled trajectory which the granule describes in a given time, Einstein considered simply its

---

* *See* LANGEVIN, *Thesis*, p. 70.

† "May soon an investigator succeed in deciding the important question in the theory of heat here proposed!"

‡ *Ann. der Physik*, 1905, 549, and 1906, 371.

displacement during this time, the displacement being defined as the length of the rectilinear segment which separates the point of departure from the point of arrival. The mean of the displacements suffered by the granule (or by a large number of identical granules) during periods of time of the same duration is the *mean displacement* appertaining to this duration.

Let us consider, provisionally, granules having the same density as the intergranular liquid; then their movement is *perfectly irregular* not only at right angles to the vertical (as under ordinary conditions) but in all senses.

In virtue of this perfect irregularity, the successive displacements distribute themselves around the mean displacement ω exactly according to the law indicated by Maxwell for the distribution of the molecular speeds around the mean speed (No. 9), and in the same way the mean square, E^2, of the displacement will be equal to $\frac{3\pi}{8} \omega^2$.

Let us suppose that the granules are unequally distributed in the liquid: they will diffuse toward the regions of lower concentration and naturally the more rapidly the more lively their movement is, that is to say, according as their mean displacement in a given time is the greater. The mathematical analysis of this idea is not very difficult * ; it implies no new hypothesis and leads to the very simple equation

$$E^2 = 6D\tau,$$

where τ indicates the duration of time considered, and D the coefficient of diffusion; this equation may be written, dividing each side by 3,

$$\xi^2 = 2D\tau,$$

ξ^2 signifying the mean square of the projection of the displacement on the axis Ox.

Let us now suppose that the granules are subjected to a force constant in magnitude and direction. Their movement, modified in the direction of this force, will not be changed at right angles to this direction and the preceding equation will still remain applicable in all that concerns the projection

* See EINSTEIN, *Ann. der Physik*, 1906, 557.

of the displacements upon a horizontal plane when the granules have not the same density as the intergranular liquid.

It remains to express the coefficient of diffusion as a function of the quantities experimentally accessible. In the case of spherical granules of radius a, Einstein arrived at this easily by considering the condition of permanent regime which is realised when a constant force*, acting upon the granules, maintains the concentrations different, in spite of diffusion, in layers perpendicular to the direction of the force. Then, by writing that as many granules pass in each instant across all planes perpendicular to the force in one sense, under the influence of the force, as pass in the other sense, under the influence of diffusion †, Einstein obtained the equation

$$D = \frac{RT}{N}\frac{1}{6\pi a \zeta};$$

but this time by the aid of hypotheses which are not necessarily implied by the irregularity of the Brownian movement.

One of these hypotheses, by which the viscosity of the liquid ζ is introduced, consisted in supposing that the law of Stokes applied in the case of granules animated by the Brownian movement. I have shown previously (No. 21) that this extension, then disputable, can be experimentally established, which relieves us from seeking a theoretical justification.

The other hypothesis, which is already familiar to us, by which Avogadro's constant N was introduced, consists in supposing that the mean energy of a granule is equal to the molecular energy. It is precisely on that account that the theory of Einstein suggests a verification of the hypothesis which attributes the origin of the Brownian movement to molecular agitation.

This verification theoretically is easy : it is sufficient to compare the two preceding equations to obtain the equation

$$\xi^2 = \tau \frac{RT}{N}\frac{1}{3\pi a \zeta},$$

---

* Which is not necessarily gravity.
† *Ann. der Physik*, 1906, 554.

in which, excepting N, only magnitudes which are directly measurable appear. It only remains to be seen whether the values of N given by this formula agree with the values otherwise found *.

Making a further step, and employing the deduction that, if the equipartition of energy is verified, the mean energy of rotation around an axis is equal to the mean energy of translation parallel to an axis, Einstein † has even succeeded in obtaining an equation which gives in a given time the mean square, a^2, of the rotation of a granule around an axis

$$a^2 = \tau \frac{RT}{N} \frac{1}{4\pi \zeta a},$$

which also can serve as a starting point of an experimental verification, more difficult but not impossible, as I shall show further on (No. 32). But first I wish to deal with those formulæ of Einstein which have reference to movements of translation.

28. Experimental proof of the theory of Einstein. First attempts.—It may first be remarked that it is the volume of the granules and not their mass which enters into this formula. Particles of dense metallic dust, droplets of oil, and even air bubbles should have thus, for equal volumes, exactly the same agitation. This is actually what good observers have already for a long time affirmed †. True these are only *impressions* not supported by any exact determinations, but they suffice at least to show that, contrary to what might have been thought, a heavy and a light granule of the same size show nearly the same agitation.

---

* It appears right to recall that Smoluchowski, almost at the same time as Einstein and by another method, arrived at a formula, but little different, in his remarkable Memoir upon a *kinetic theory of Brownian movement* (*Bulletin de l'Académie des Sciences de Cracovie*, July 1906, translated into French), where is to be found, in addition to some very nteresting reflections, an excellent historical summary of work previous to 1905.

† *Ann. der Physik*, 1906, 371-380.

† JEVONS, *Proc. Manchester Phil. Soc.*, 1869, 78. CARBONELLE and THIRION, *Revue des Questions scientifiques*, 1880, 5. GOUY, *Comptes rendus*, 1889, cix. 102.

Beyond the size of the granule, the theory of Einstein does not take into account the electrification which in general occurs at the surface of contact of a liquid. The contrary has been supposed by various authors who affirm, without otherwise assigning the reason, that the electrification of the granules was the condition necessary for their agitation. The error of this hypothesis has been shown by Svedberg [*], who, by gradually adding traces of sulphate of aluminium to a colloidal solution of silver, reversed the sign of electrification of the granules and passed through a zero value of this electrification without perceiving at any instant the least abatement in the activity of the Brownian movement.

As regards the influence of temperature and viscosity some interesting observations of Exner may first be cited, although their significance indeed may be uncertain, but being anterior to the theory of Einstein, they could not in any case have been influenced by this theory [†].

Exner worked on granules of gamboge, of which he estimated the radius (correct to about 25 per cent. ?) according to the appearance of the image (enlarged as is known by diffraction). He followed as well as possible in the *camera lucida* the trajectory during a given time, and dividing by this time the total curvilinear path so obtained, hoped to arrive at the true speed of the granule, at least approximately. We have seen (No. 13) that such estimates are altogether false and that the true speed is enormously greater than the apparent speed so obtained. But the ratio, at two temperatures, of the lengths of trajectories delineated during a given time may not be (?) very different from the ratio of the lengths of chords joining the extremities of the trajectories. In other words, the ratio of the supposed *speeds* observed by Exner at two temperatures may be approximately equal to the ratio of the displacements during the same time for these two temperatures.

Now Exner states that the *speed* of a granule is multiplied by about 1·6, when the temperature is raised from 20° to 71°. This number is nearly equal to 1·7, the square root of the

---

[*] *Nova Acta Soc. Sc.*, ii. Upsala, 1906.
[†] E. EXNER, *Ann. der Physik*, 1900, ii. 843.

ratio $\dfrac{273+71}{273+20} \cdot \dfrac{0\cdot010}{0\cdot004}$, which, according to Einstein, ought to be the square of the ratio of the mean displacements during the same time at the two temperatures considered. It can be seen that there is here at least a presumption of a partial verification of the formula in question.

Incidentally, it can be seen from this example that the activity of the Brownian movement which accompanies a variation of temperature depends above all on the correlative variation of viscosity. Exner, who hoped to arrive at the true speeds, considered that what he believed to be the kinetic energy of the granule was very far from varying proportionally to the absolute temperature, and concluded wrongly that the granules could not be regarded as analogous to the molecules of a fluid.

Some years later, and this time in possession of the formula of Einstein, Svedberg attempted at once an experimental control and thought he had obtained a satisfactory verification *. But I must say that this part of his work, very interesting in other respects †, does not appear to me to justify the optimistic conclusions he draws from them, and leaves the question proposed without answer. The displacements which he has observed are from 4 to 6 times larger than what according to his calculations should verify the formula; at first sight, taking into account the experimental difficulties, there is a temptation to see there at least a rough agreement, but a careful examination discloses a discrepancy actually enormous. Actually a displacement 4 to 6 times too large requires, if the formula is exact, a radius 25 times smaller than the radius assumed by Svedberg, that is to say, for the same weight of substance about 12,000 times more granules in a given volume. Now one of the methods employed by Svedberg to find the radius consisted precisely in the counting of the number of granules in a known volume

---

* *Studien zur Lehre von den kolloidalen Losungen* (*Nova Acta Reg. Soc. Sc. Upsaliensis*, 4th series, ii., Upsala 1907), and "Ion," 1909.

† The absence, referred to above, of all relation between the contact electrification and the Brownian movement, and the discovery of colloidal metallic solutions in non-ionised liquids.

of titrated emulsion, and it is quite impossible that he should have seen in this volume 12,000 times fewer granules than it contained. But further : Svedberg, to make his calculations, attributed to Avogadro's constant N the value 4.10^{23}, which was admissible at that time, but which is certainly too little by almost one-half. Giving N a more exact value it will be seen that the mean displacements he indicates are more than 7 times too great, and it is necessary that he should have found in the volume explored 125,000 times too few granules. The obvious conclusion from the experiments of Svedberg would thus be, contrary to what he says, that the formula of Einstein is certainly false.

Fortunately there is probably little in common between the magnitudes ξ and τ which enter into the formula and the ill-defined magnitudes introduced in their place by Svedberg. By the aid of a suitable flow he impressed upon the emulsion a uniform movement so rapid that, on account of the persistence of luminous impressions, each granule would give to the eye a brilliant trajectory. By virtue of the Brownian movement this curve is jagged at right angles to the displacement of the whole. But from what is known of the absolute irregularity of the Brownian movement, it would appear quite impossible that these trajectories can be, as Svedberg, evidently the victim of an illusion, describes them, " lines regularly undulated, of well-defined amplitude and wave-length ! " By comparison with a micrometer eye-piece, Svedberg *estimated* the magnitude of the quantities (in reality non-existent) which he calls *the wave-length* and the *amplitude of oscillation*. Taking into account the speed of flow of the liquid he calculated then the time *of duration of the oscillation*, which will be the time during which a granule suffers, at right angles to the displacement of the whole, a displacement equal to double the amplitude. I do not think it necessary to insist upon the uncertainty, to my mind *complete*, which results from a method so questionable and from estimations so vague. On the contrary, it is *a priori* completely correct to obtain the mean displacement of a granule in a fixed time by dotting the successive positions of the image of this granule on photographs taken at equal intervals of time. Victor Henri has made in this sense a cinematographic study of the Brownian movement of the natural granules of the latex of Caoutchouc. He worked with relatively large granules, the diameter of which was estimated, from the size of the images, to be about 1 μ.*

Except that the mean displacement in a given time varied very nearly as the square root of the time, the whole of these measurements appeared unfavourable to the theory of Einstein. In neutral water the mean displacement, being nearly three times greater than the formula would indicate, can only agree with it if the diameter of the granule evaluated at 1 μ, was in reality 8 times smaller; now this is not admissible (at this diameter the granule would not even be visible, the illumination being direct).

But, more serious still, it seemed that traces of acid or alkali, which do not appreciably change the viscosity, and besides are insufficient to coagulate the caoutchouc granules, abate their movement very markedly. For example, in feebly acidulated water, the mean displacement became about 9 times less than in neutral water, which requires for granules of the same external appearance a diameter 80 times greater than in neutral water. This enormous variation is completely irreconcilable with the theory of Einstein, and, in general, with all theories which neglect the nature of the intergranular fluid and only make it intervene by its viscosity.

As far as I can judge by conversation, this then produced, among the French physicists who closely follow these questions, a current of opinion which struck me very forcibly as proving how limited, at bottom, is the belief we accord to theories, and to what a point we see in them instruments of discovery rather than of veritable *demonstrations*.

Without hesitation it was supposed that the theory of Einstein was incomplete or inexact. On the other hand, molecular agitation could not be abandoned as the origin of the Brownian movement, since I had shown by experiment † that a dilute emulsion behaves as a very dense perfect gas, of which the molecules have a weight equal to that of the

---

* *Comptes rendus*, 1908, 18th May and 6th July.
† *Comptes rendus*, 1908, cxlvi. 967.

granules of the emulsion. It had therefore to be supposed that some unjustifiable complementary hypothesis had entered unperceived into the reasoning of Einstein.

29. Experimental confirmation of Einstein's theory.

—However, as Victor Henri had only estimated the diameter of his granules, and as he himself had made some reservations as to the generality of his results, I thought it would still be of use to measure the mean displacement of the granules of exactly known diameter which I knew how to prepare. A student who worked in my laboratory, M. Chaudesaigues, agreed to take charge of these measurements *. In default of a chronophotographic apparatus, he dotted the position of a granule in the *camera lucida* from half-minute to half minute, recommencing with another granule, and so on, making in general four readings on each granule.

Right from the first measurements it became manifest, contrary to what might have been expected, that the displacements verified at least approximately the equation of Einstein. At the same time I satisfied myself that the addition of traces of acid did not appreciably alter the movement of the granules, provided that these granules were remote from the walls †. In brief, it was necessary to suppose that some unknown complication, or some source of systematic error, had vitiated the results of Victor Henri, for the measurements, of which I am about to give a summary, cannot leave any doubt of the rigorous exactitude of the formula proposed by Einstein.

As I said, the granules which I had prepared were dotted in a *camera lucida*, the microscope being vertical, which gives the horizontal projection of the displacements. Working on squared paper, the projection on two rectangular axes of the different segments are so obtained directly, but it is useless to measure them, for the sum of the squares of these projections is equal to the sum of the squares of the segments,

---

* *Comptes rendus*, 1908, cxlvii. 1044, and *diplome d'Etudes*, Paris, 1909.

† Although this addition annulled and then changed the sign of the electrification, which these granules assume by contact with water.

whence it follows that, to obtain the mean square of the projection upon an axis, it is sufficient to measure these segments one by one, to calculate their squares, and to take the half of the mean of these squares. It only remains then to see whether the value given for N by the equation of Einstein,

$$\xi^2 = \tau \frac{RT}{N} \frac{1}{3\pi a \zeta},$$

agrees, within the error of experiment, with the value already determined.

In a preliminary trial, M. Chaudesaigues studied some relatively large granules of gamboge, of radius about $0{\cdot}45\ \mu$, which were moderately uniform. He noted the displacement of 40 of these granules during one minute and of 25 during two minutes; these positions gave for N the value 94.10^{22}. On the other hand, 30 granules practically identical, of a slightly greater radius, equal to $0{\cdot}50\ \mu$, gave me 66.10^{22}, which makes a mean of 80.10^{22} for this group of granules.

M. Chaudesaigues then studied the granules of radius equal to $0{\cdot}212\ \mu$, which had served for my most exact determination of N (No. 24). The two tables following summarise the measurements made with two series of 50 granules, following each from 30 seconds to 30 seconds for two minutes, the viscosity being 0·011 for the first series (water at 17°) and 0·012 for the second :—

First Series.

Time in seconds.	Mean Horizontal Displacement (in μ).	$\xi^2 \times 10^{-8}$.	$N \times 10^{-22}$.	N (mean).
30	8·9	50·2	66	
60	13·4	113·5	59	73×10^{22}
90	14·2	1 8	78	
120	15·2	144	89	

Second Series.

Time in seconds.	Mean Horizontal Displacement (in μ).	$\xi^2 \times 10^{-8}$.	$N \times 10^{-22}$.	N (mean).
30	8·4	45	68	
60	11·6	86·5	70·5	68×10^{22}
90	14·8	140	71	
120	17·5	195	62	

Lastly, in a third series, always with granules of the same radius, the intergranular liquid was water containing much sugar, nearly five times more viscous than pure water. The mean displacement in 30 seconds, now equal to 4·7 μ, is so reduced nearly in the predicted ratio (within about 10 per cent.) and gives for N the value 56×10^{22}, rather lower than the preceding, without, however, the divergence exceeding what is possible by reason of the irregularities of statistical values, and of sources of error generally, which are a little augmented by the greater complication of the experiment.

The rough mean of these four series of measurements, which is practically equal to the mean of the two best series taken alone, is exactly

$$70 \cdot 10^{22},$$

and is practically identical with that found by the completely different method, based upon the distribution of the granules in permanent regime. The arrangement is as perfect as possible and, to reiterate, no doubt can remain.

If, to the different values indicated for N in the preceding Tables, is attributed a *weight* proportional to the number of determinations which gave them (more numerous, for example, for the interval of 30 seconds than for that of 120 seconds) a slightly different general mean will be obtained, namely 68·7 (not 64 as erroneously published in the *Comptes rendus*) instead of 70. I do not think it necessary to make this small correction, because of a source of error, the explanation of which is of some interest, which is of greater importance for the short intervals of time than for the long. Each time a granule is dotted, a small error is really made, analogous to that in target-shooting, which obeys the laws of chance and has the same effect on the readings as if a second Brownian movement were superimposed upon that which it is desired to observe. The corresponding error, which of necessity causes an increase in the calculated mean square ξ^2, while insignificant for large intervals of time and for small viscosities, becomes of the greater importance as the time-interval is reduced and the viscosity increased. It will always have the effect of diminishing slightly the value which would be given for N by rigorously exact dotting.

30. Second confirmation of Einstein's formula.—It was desirable to check these results by changing the substance employed and, as for the distribution in height, with the collaboration of M. Dabrowski, I have repeated the measurements, substituting mastic for gamboge.

The granules of the uniform emulsion studied had a radius equal to $0{\cdot}52\,\mu$. The illumination was provided by an Auer light and, as in the preceding experiments, the beam traversed a thick cell full of water, which stopped almost all the rays capable of warming the water of the preparation. This was immersed in water, so that the objective was used with immersion, and from time to time the temperature (which it is important to know correctly, because of its great influence on the viscosity) was measured by introducing a thermometer into the tube of the microscope in contact with the objective.

We first made two series of measurements, taking turns at the microscope, each dotting the granules every 30 seconds at the call of the other. In each series this interval of 30 seconds corresponded to about 200 points, the interval of 60 seconds to 100 points, and so on. The results are summarised in the following Table :—

Time in seconds.	N×10⁻²².	
	First Series.	Second Series.
30	57	69
60	64	65
120	67	64
240	70	88

Lastly, in a third series, we have measured for 200 distinct granules the magnitude of the displacement in 2 minutes. These 200 measurements have given for N the value

$$77 \cdot 10^{22}.$$

These different measurements indicate for N a mean value comprised, according to the widest conventions that can be made as to the relative importance of the numbers dotted, between $72 \cdot 10^{22}$ and $74 \cdot 10^{22}$, and so is about

$$73 \cdot 10^{22}.$$

Taking into account the measurements already made with gamboge, it can be said that the 3000 displacements together indicate for $N \cdot 10^{-22}$ the value

71·5,

which agrees well with the value 70·5 (probably a little nearer) obtained by the fundamentally different method I used first.

The figure here reproduced (fig. 6, p. 64) shows three drawings obtained by tracing the segments which join the consecutive positions of the same granules of mastic at intervals of 30 seconds. It is the half of the mean square of such segments which verifies the formula of Einstein. One of these drawings shows 50 consecutive positions of the same granule. They only give a very feeble idea of the prodigiously entangled character of the real trajectory. If the positions were indicated from second to second, each of these rectilinear segments would be replaced by a polygonal contour of 30 sides, relatively as complicated as the drawing here reproduced, and so on. One realises from such examples how near the mathematicians are to the truth in refusing, by a logical instinct, to admit the pretended

Fig. 6.

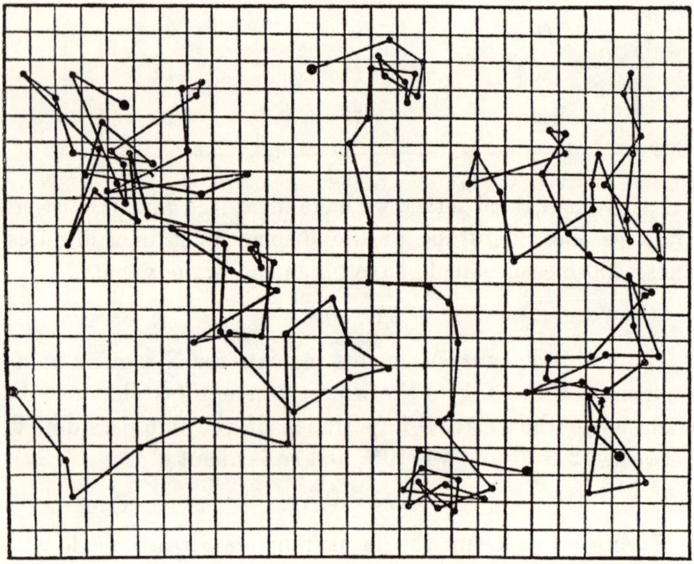

geometrical demonstrations, which are regarded as experimental evidence for the existence of a tangent at each point of a curve.

31. The law of distribution of the displacements.—We have shown (No. 27) that, in the case of granules having the density of the intergranular liquid, the displacements in a given time ought to distribute themselves around the mean displacement according to the law of irregularity of Maxwell (No. 9). It is useful to verify directly this important law. This can be done in various ways.

First the probability that the component along the axis Ox should be comprised between x and $x+dx$ should be

$$\frac{1}{\xi} \frac{1}{\sqrt{2\pi}} e^{-\frac{x^2}{2\xi^2}} dx,$$

designating always by ξ^2 the mean square of the component x, a result which will hold valid for all horizontal axes when the granules no longer have the density of the intergranular liquid (No. 27). Of $\mathfrak{N}$ observations the number which will give components lying between x_1 and x_2 will then be calculable from the expression

$$\mathfrak{N} \; \frac{1}{\xi} \frac{1}{\sqrt{2\pi}} \int_{x_1}^{x_2} e^{-\frac{x^2}{2\xi^2}} dx.$$

M. Chaudesaigues has made this calculation, relatively to an arbitrary horizontal axis, for the displacements suffered in 30 seconds by the granules of *gamboge* (Tables of No. 29). The numbers n of displacements having their projection comprised within two given limits (multiples of $1.7\,\mu$ which correspond to 5 mm. of the squared paper) are indicated in the following Table :—

Projections (in μ) comprised between	First Series.		Second Series.	
	n (found).	n (calc.).	n (found).	n (calc.).
0 and 1·7	38	48	48	44
1·7 ,, 3·4	44	43	38	40
3·4 ,, 5·1	33	40	36	35
5·1 ,, 6·8	33	30	29	28
6·8 ,, 8·5	35	23	16	21
8·5 ,, 10·2	11	16	15	15
10·2 ,, 11·9	14	11	8	10
11·9 ,, 13·6	6	6	7	5
13·6 ,, 15·3	5	4	4	4
15·3 ,, 17·0	2	2	4	2

Another verification, perhaps still more striking, the idea of which I owe to Langevin, consists in transporting parallel to themselves the observed horizontal displacements, in such a manner as to give them all a common origin. The extremities of the vectors so obtained should distribute themselves around this origin as bullets fired at a target distribute themselves around the bull's-eye. This is well seen in fig. 7, where I have recorded 365 observations relating to granules of mastic, of which I have spoken in the preceding paragraph. Here again the checking of the law of distribution can be

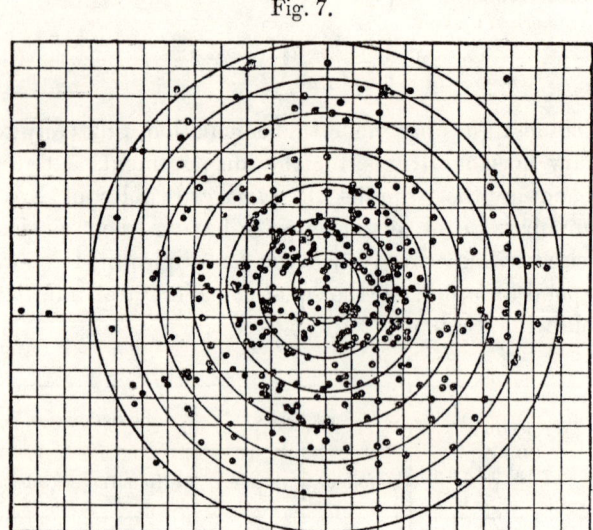

Fig. 7.

quantitative. For if the law of probability given for a component x is admitted, it is easy to see that the probability of a horizontal displacement having a length comprised between r and $r+dr$ is given by the expression

$$\frac{1}{2\pi\xi^2}e^{-\frac{r^2}{2\xi^2}}2\pi r\, dr\,;$$

or, simplifying and replacing $2\xi^2$ by the mean square, ρ^2, of the horizontal displacement,

$$\frac{2}{\rho^2}e^{-\frac{r^2}{\rho^2}}r\,.\,dr$$

of which the integral is simply $-e^{-\frac{r^2}{\rho^2}}$, so that it follows that the number of displacements comprised between r_1 and r_2 can be immediately calculated.

In the case of the preceding figure, ρ is equal to $7\cdot16\,\mu$, and I find for the number of displacements comprised between two fixed limits :—

Displacements (in μ) comprised between	n observed.	n calculated.
0 and 2	24	27
2 and 4	76	71
4 and 6	90	84
6 and 8	67	76
8 and 10	45	54
10 and 12	34	30
12 and 14	20	14
14 and 16	4	5
16 and ∞	5	4

I find, lastly, a third form of verification in accord with the value actually found for the mean square of the horizontal displacement ρ^2 and with that which can be assigned to it as soon as the mean value ϵ of the horizontal displacements is known.

The reasoning is very analogous to that which permits it to be shown, assuming the law of Maxwell, that the mean square of the speed U^2 is obtained by multiplying by $\dfrac{3\pi}{8}$ the square Ω^2 of the mean speed (No. 9). Let us indicate this reasoning.

We see that out of $\mathfrak{N}$ displacements, there are between r and $r + dr$

$$\mathfrak{N}\frac{2}{\rho^2}e^{-\frac{r^2}{\rho^2}}r\,.\,dr,$$

which have for the sum of their lengths

$$\mathfrak{N}\frac{2}{\rho^2}e^{-\frac{r^2}{\rho^2}}r^2\,.\,dr.$$

The sum of all the lengths is thus

$$\mathfrak{N}\frac{2}{\rho^2}\int_0^\infty \epsilon^{-\frac{r^2}{\rho^2}}r^2\,dr,$$

which is

$$\mathfrak{N}\frac{2}{\rho^2}\frac{\sqrt{\pi}}{4}\rho^3 = \mathfrak{N}\rho\sqrt{\frac{\pi}{4}},$$

and the mean length ϵ of the displacement which is obtained by dividing this sum by the total number $\mathfrak{N}$ of the displacements is thus

$$\epsilon = \rho\sqrt{\frac{\pi}{4}},$$

that is to say, very nearly,

$$\rho = \frac{9}{8}\epsilon.$$

This result is well verified. For example, the displacements which have served for the preceding figure have a mean equal to 6·4 μ. The value predicted for ρ by this calculation is hence

$$7\cdot 21\,\mu,$$

which is in good agreement with the value

$$7\cdot 16\,\mu$$

actually found.

In brief, Maxwell's law of irregularity is verified indisputably in its application to the displacement of the granules of an emulsion.

That shows that the probability of a certain value x of the projection of a displacement on the axis Ox does not depend upon the values of the components y and z (No. 9). After this it is difficult to doubt the independence of the three components of the speed. It comes to the same thing to say that it is now difficult to doubt Maxwell's law of distribution of velocities, although the completely direct verification, realised here for the displacements, is still wanting for the velocities.

32. Special study of very large granules.—The preceding reasoning does not contain any restriction as to the size of the granules, and, as regards this, seems to assume simply that the mass of the fluid can be considered very large relatively to the mass of the granules, so enabling the influence of the boundaries to be neglected. So that if no hypothesis has escaped us, objects of large size have still a perceptible Brownian movement, since according to the formula of Einstein a ball of 1 mm. in diameter should have in water at 20° an agitation corresponding to a mean displacement of 1 μ each minute.

Without being able to experiment upon such large particles, I have, however, considerably extended the region in which certain verification of the preceding laws can be obtained.

It was first necessary *to prepare* at will spherical granules considerably larger than those of the emulsions so far studied. I have succeeded in doing this in a manner which the following reasoning will elucidate.

When, without precautions, some water is poured into an alcoholic solution of resin, an aqueous solution is suddenly formed which is very strongly supersaturated with resin at every point. If what is known in other directions of the spontaneous separation of an unstable phase into two phases (formation of small drops, precipitation of crystals) be borne in mind, it is not surprising that the *germs* around which the insoluble phase will grow, appear to be in extremely great numbers. Each of these can only exhaust a very small space and will give therefore a very small granule, so that actually, when an alcoholic solution is violently mixed with water, the granules produced have a diameter in general far below 1 μ. But if, instead of using pure water, the alcoholic solution is mixed with water containing much alcohol, in which the solubility of the resin, although small, is still appreciable, there will be a chance of the germs being produced in much smaller numbers, and in consequence the spheres of resin, which can only form around such germs, may be much larger than in the preceding case.

This, in reality, I have effected by allowing pure water to flow slowly out of a funnel with a narrow orifice under an alcoholic solution of about 5 per cent. of gamboge or mastic. A zone of continuous passage then, necessarily, is established between the two liquids : as soon as a layer contains sufficient water for the supersaturation to be considerable there, germs form, which grow by intercepting the resin arriving from the upper layers, without the supersaturation ever rising to such a degree as will bring about the appearance of very numerous germs. Finally, by reason of this growth the granules soon become so heavy that they fall, in spite of their Brownian movement, passing through the lower layers of pure water, where they are washed, to the bottom of the vessel, and it only remains to collect them by decantation. I have thus precipitated all the resin out of alcoholic solutions of gamboge or mastic in the form of spheres, of which the diameter, practically never less than $2\,\mu$, is generally in the neighbourhood of $10\,\mu$ and can attain $50\,\mu$. These large spheres have the appearance of glass balls, yellow for gamboge, colourless for mastic, which are easily broken into irregular fragments: they frequently appear perfect and give, in the same way as lenses, a recognisable real image of the luminous source which illuminates the preparation (for example an Auer mantle) upon which it is easy to focus the microscope. But nearly as often they contain inclusions of slightly different refractivity. I have not been able to settle completely the origin and nature of these inclusions * : thanks to them it is easy to perceive the irregular movements of rotation of the spheres.

Lastly, in exceptional fashion it happens that a granule may be formed of two spheres joined together all round a little circle, resulting evidently from the welding of two spheres while they were in progress of growth around their respective germs. From the double point of view of the origin of *germs* and of their speed of growth, these various appearances present some interest beyond the special end here pursued.

* These inclusions cannot have a composition very different from that of the rest of the granules, for they scarcely modify its density, as proved by the method of Retgers of floating them in an aqueous solution of urea. I think that they are formed by a very viscous mixture enclosing a little alcohol, similar to that which slowly separates when an alcoholic solution already containing a little water is diluted with a very little water (the next day there is found a thin, very viscous layer of almost pure resin at the bottom of the vessel). Some drops of this nature, coming by Brownian movement into a layer where the vitreous resin, properly so-called, is separating can be united in the more rapid growth of the spheres of this pure resin (the reason of this rapidity being unknown to me).

However it may be with regard to these new problems, we now know how to prepare large spherical granules. They can be separated according to size by a fractionation similar to that successfully used for the small granules (No. 16), but in this case centrifuging is not necessary, and the separation is simply accomplished by making use of the fact that the largest granules fall the most rapidly. Let us see how the fundamental laws of the Brownian movement can be followed for such granules having, for example, a diameter of 10 to 12 μ.

One can scarcely hope to study in water their progressive rarefaction as a function of the height. It suffices to apply the formula of rarefaction (No. 24) to perceive that in the case of mastic each elevation of only 1 μ suffices to divide the concentration of the granules by about 60,000 (the rarefaction will be still more rapid with gamboge). This comes to saying that all the granules are assembled together in the immediate neigbourhood of the bottom, which can actually be verified, but which does not permit of measurement.

On the contrary, the granules distribute themselves throughout the depths of the preparation if a density practically equal to that of the granules is given to the intergranular liquid by dissolving therein a suitable substance. Even then, the quantitative verification of the law of distribution still remains practically impossible, for to make measurements approximate to a hundredth part, it would be necessary to determine the densities to a millionth.

But the measurement of the mean displacement does not appear, at least *a priori*, as if it would present any serious difficulty, and one can try to see if the formula of Einstein still applies.

I have therefore added different substances to the intergranular liquid, so as to give it the same density as the granules. A complication soon became manifest, for the greater number of these substances *coagulate* the granules, showing in addition in the most happy manner in what the phenomenon of **coagulation** consists, which is less easy to comprehend for ordinary colloidal solutions with ultra-microscopic granules. Under the influence of the coagulant the large granules studied are seen to arrange themselves in clusters of conjoined granules, rather like bunches of grapes, or even like regular pilings of bullets.

This simple and direct result divests of probability the complicated hypothesis, permissible so long as we cannot *see* the phenomenon of coagulation in detail, according to which the granules of a coagulum can be connected the one to the other without being in contact.

In quantity necessary to cause the granules to float in the middle of the liquid, the salts ordinarily considered as feeble coagulants, and even sugar, all coagulated my large granules of mastic. Urea alone had a coagulating power weak enough to allow the movements of single granules to be followed.

The intergranular liquid of suitable density contained about 27 per cent. of urea and its viscosity was 1·28 times that of pure water. A portion of the granules then floated between two layers of liquid and could be usefully observed : their number was always very small, since very feeble differences of density, it is easy to see, are sufficient to bring all the granules close to the bottom or to the surface. Certain measurements were made in a cell 1 mm. high, as high relatively to the large granules as the ones first used were to the small. But on examination the movements appeared the same in this high cell as in cells of only $100\,\mu$; and it is to be supposed that the distribution of molecular movements around the granule which determines its movement, attains its normal regime at a distance from the walls which has nothing to do with the size of the granules. I have followed in the *camera lucida* two of these granules which were practically equal (of diameter $11\cdot5\,\mu$), and have measured about 100 displacements. They gave for N by the application of Einstein's formula $78\cdot10^{22}$. In other words the mean horizontal displacement per minute at 25° was found equal to

$$2\cdot35\,\mu,$$

whereas the calculated value ($N = 70\cdot5 \times 10^{22}$) would be

$$2\cdot50\,\mu.$$

Bearing in mind all the difficulties encountered and the

small number of points, the agreement is almost unexpected, and there is no doubt that Einstein's theory remains valid.

Now, the fundamental principle of this theory is the *theorem of the equipartition of kinetic energy*. It is therefore established by the preceding experiments that a granule of mastic of 11·5 μ diameter has the same mean kinetic energy as the smallest granules studied in my other experiments *which weigh about* 60,000 *times less. In this, I think, will be found much the most extended verification up to this day of the equipartition of the kinetic energy of translation.*

33. The Brownian movements of rotation and the energy of rotation.—Lastly, thanks to the inclusions which render manifest a spontaneous irregular rotation of large spherical granules, I have been able to establish by experiment one of the most important propositions of the kinetic theory, that is to say : the mean equality of the energies of rotation and translation. This proposition enables the equation given by Einstein for the rotations to be established :

$$\alpha^2 = \tau \, \frac{RT}{N} \, \frac{1}{4\pi\zeta a^3},$$

and it is sufficient to see whether this equation is verified.

In fact, although the existence of a lively Brownian movement of rotation has often been recorded, no one has ever tried to measure it, which can be understood when it is observed that if the formula is exact, a mean rotation of 100° per second is indicated for granules of 1 μ in diameter. But for my granules of 10 μ to 15 μ the predicted rotation is no more than some degrees per minute and should be easily measurable. I have, indeed, succeeded in fixing from minute to minute the orientation of spheres of mastic having a diameter of about 13 μ in suspension in a solution of urea : it is sufficient for this purpose to dot the successive positions of small inclusions, the distances of which from the centre have been measured. Then the necessary elements for calculating the component of the rotation around any axis are known. The numerical calculations, of which the details are of no interest, give for N, after making use of about 200 angular measurements, the value $65 \cdot 10^{22}$. In other words, these measurements indicate for $\sqrt{\overline{\alpha^2}}$ per minute the value

$$14°\cdot 5,$$

whilst the calculation predicts a rotation of

$$14°.$$

The agreement is remarkable, if one thinks of the difficulties of measurement and the complete uncertainty which *a priori* surrounded even the order of magnitude of the rotation. The granules utilised for these measurements were about 100,000 times heavier than the granules of gamboge first studied.

So *the equipartition of energy* is established throughout this great interval. Incidentally, its verification for the rotations is an experimental confirmation of the reasoning from the kinetic theory which has enabled the ratio $\dfrac{C}{c}$ of the specific heats of a perfect gas to be predicted.

34. Recapitulating, the molecular kinetic theory of Brownian movement has been verified to such a point in all its consequences that, whatever prepossession may exist against Atomism, it becomes difficult to reject the theory. In the second place, the quantitative study of the law of distribution of the granules of an emulsion on the one hand, and, on the other, of the activity of Brownian movement, leads, in two different ways, to *exactly the same value* for Avogadro's constant, which is essentially invariable on the kinetic theory.

It is interesting to compare this value with those obtained in other ways. Although, for the most part, they do not yet allow of any precision, their agreement assumes great importance, as demonstrating the extreme diversity of the methods which have furnished them.

Without being able to explain these methods in detail, I wish at least to enumerate them, so as to facilitate a just perspective of the whole of the questions, in which molecular reality imposes itself on the attention most forcibly.

IV.

35. Indications given by diffusion.—As this point of view is immediately related to the theory of Einstein, I will first say a few words in general as to the somewhat vague information which can be obtained from the measurement of the coefficient of diffusion.

According to one of the formulæ of Einstein the coefficient of diffusion of spherical granules is given by the expression

$$\frac{RT}{N} \frac{1}{6\pi\zeta a}.$$

It may be hoped that this formula still applies roughly to the case of molecules as small as sugar or phenol, and thus it may be seen whether the formula, so assumed, leads to acceptable values of N. This is naturally what Einstein [*] tried as soon as he was in possession of this formula.

In the case of sugar at 18° one should have approximately

$$\frac{0\cdot 33}{86,400} = \frac{83\cdot 2 \times 10^6 \times 291}{6\pi \times 0\cdot 0105} \frac{1}{Na},$$

or

$$Na = 3 \cdot 10^{-16}.$$

It remains to find the radius (?) of the sugar molecule. The simplest plan is to consider it as approximately given by the specific volume of solid sugar (Langevin), or, in a manner a little more precise still, to observe that in the solid the molecules cannot be packed closer than in a pile of bullets (No. 11), and are probably less scattered than in an ordinary liquid (where, according to Van der Waals, the apparent volume is four times the real volume of the molecules).

There results for N a value comprised between

$$85 \cdot 10^{22} \quad \text{and} \quad 150 \cdot 10^{22}.$$

The same calculation applied to phenol, for which the structural formula indicates a very compact molecule, gives for N a better value, comprised within

$$60 \cdot 10^{22} \quad \text{and} \quad 100 \cdot 10^{22}.$$

[*] *Ann. der Physik*, 1906, xix. 289.

In reality Einstein obtained the radii by a more complicated and uncertain process, according to the difference between the viscosities of pure water and the solution (*Ann. der Physik*, 1905). He so found in the case of sugar a value of N equal to

$$40 \cdot 10^{22}.$$

All these values are roughly concordant, and better can scarcely be hoped from a line of reasoning which supposes the molecules of saccharose to be spherical (it is much more probable that they resemble long cylinders). For the cyclic molecule of phenol the result is already much better and practically as near as the result obtained by application of the theory of Van der Waals (No. 11).

Now the reasoning of Einstein supposes the *law of Stokes* to be valid. It is therefore probable that this law, the exactitude of which I have proved directly as far as dimensions of the order of a tenth of a micron (No. 21), *still remains exactly verified for large molecules, the diameter of which does not reach the thousandth of a micron.* This without doubt is the most interesting result which we owe to this consideration of the coefficients of diffusion. It will permit us presently to apply the law of Stokes with safety to the case of ions in movement through a gas (No. 38).

36. Indications given by the mobility of ions (in liquids).— A still more daring extension of the law of Stokes is at the bottom of a very ingenious idea developed by M. Pellat[*]. Let v be the mean speed of electric transport, shown by a monovalent ion of radius a and charge e in an electric field H. We shall have, if the law of Stokes is applicable,

$$6\pi\zeta av = \mathrm{H}e,$$

an equation which finds at least a partial verification in the well-known fact that the speed v is proportional to the field. Let us multiply the two sides of this equation by Avogadro's

---

[*] *Traité d'Électricité*, vol. iii. p. 56.

constant N; we shall have, remembering that Ne has the value 29×10^{13} electrostatic units,

$$6\pi\zeta\frac{v}{H} aN = 29 \cdot 10^{13}.$$

If, on the other hand, we suppose that the volume of the *charged* ion may be approximately calculated, starting from the volume Φ of the gram-atom in the solid state, in the same way as that of a neutral molecule may be calculated from the gram-molecule of the solid (an hypothesis which will appear possible but not necessary, if the manner in which the radius of an atom is defined by the impacts (No. 10) is borne in mind) we shall have, approximately, as indicated in the preceding paragraph,

$$0.25\, \Phi < \tfrac{4}{3}\pi a^3 N < 0.73\, \Phi,$$

and these two relations will furnish an order of magnitude for a and N.

Let us apply this, not to mercury as M. Pellat did (for the mobility of the mercury ion is unknown and can only be fixed in a hypothetical manner), but to the monovalent ions which are the best studied.

For silver, the atomic volume of which is low, we have

$$63 \cdot 10^{22} < N < 108 \cdot 10^{22},$$

giving, let us say, a mean value of $85 \cdot 10^{22}$, practically as near as the values given by Van der Waals.

But the alkali-metals, and especially cæsium, which, while possessing in the state of ions a mobility not greatly different from that of silver, have a much higher atomic volume (which, besides, is very well known), give much less satisfactory values for N. Thus, to 30 per cent. more or less, the potassium ion will give the value $30 \cdot 10^{22}$ and the cæsium ion $15 \cdot 10^{22}$, almost three times too low. Such a disagreement would not have seemed very large only a few years ago, and no one would have dared to affirm even that there was a certain disagreement. Now we have the right to say that this result shows, either that the law of Stokes begins decidedly no longer to apply at this extreme degree of smallness (without, however, the deviation becoming yet very great), or that the atom of potassium, for example, has a radius two and a half times smaller when it is in water, in the form of an ion, than when it is in the solid metal.

It would be necessary to obtain a second relation, got without reference to the law of Stokes, to elucidate this question.

37. Indications drawn from the blue colour of the sky.—A very curious and completely different method, due to Lord Rayleigh, brings in the diffraction of the light which comes from the sun by the molecules of the atmosphere.

When a pencil of white light penetrates a medium where fine dust-particles are present, the trajectory of the ray is rendered visible laterally, owing to the light diffused or diffracted by these particles. The phenomenon persists when the particles become more and more fine (and it is this which makes possible *ultramicroscopic* observation), but the opalescent light diffracted tends to become blue, the light of short wave-length thus suffering the more pronounced diffraction. Further, the light so scattered is found to be polarised in the plane passing through the incident ray and the eye of the observer.

No limit of minuteness is *a priori* assigned to the diffracting particles. Lord Rayleigh supposes that even the molecules act like the particles still visible in the microscope, and that this is the origin of the blue light which comes to us from the sky during the day. In accord with this hypothesis the blue light of the sky, observed in a direction perpendicular to the solar rays, is strongly polarised. It is, in addition, difficult to suppose that this is due to a diffraction by dust-particles properly so-called, for the *blue of the sky* is scarcely enfeebled when we ascend 2000 to 3000 metres in the purest atmosphere, well above the greater part of the dust-particles which sully the air in the immediate neighbourhood of the soil.

Without resting content with this qualitative conception, Lord Rayleigh, developing the elastic theory of light, has calculated the ratio which should exist, according to his hypothesis, between the intensity of the direct solar radiation and that of the blue light. In a precise manner, let us suppose that the sky is observed in a direction making an angle ϕ with the vertical and an angle β with the solar rays; the illuminations e and E obtained at the focus of an objective, successively directed towards this region of the sky and towards the sun, are for each wave-length λ in the ratio

$$\frac{e}{\mathrm{E}} = \left(9\pi^3\omega^2 \frac{1+\cos^2\beta}{2\cos\phi}\right) \frac{p}{\mathrm{M}g} \frac{\mathscr{R}^2}{\lambda^4} \frac{1}{\mathrm{N}},$$

ω indicating the apparent semi-diameter of the sun, p the atmospheric pressure at the place of observation, g the acceleration of gravity, M the mass of the gram-molecule of air, $\mathscr{R}$ the molecular refractive power of air

$$\left(\text{that is } \frac{\mathrm{M}}{d} \frac{n^2-1}{n^2+2}\right)$$

and N Avogadro's constant, which can thus be fixed by this equation, supposing it to be exact. The probability of this exactitude is, besides, increased by the fact that Langevin, starting from the electromagnetic theory, obtained exactly the same equation (n^2 being replaced by the dielectric constant K). It will be seen that the extreme violet of the spectrum suffers a diffraction about sixteen times greater than the extreme red, which well explains the observed colour.

To be exact a test of this theory should be carried out at a height sufficient to avoid disturbances due to *dust-particles* (fumes, mists, *large ions*, etc.). Further, the measurements ought to be spectro-photometric. This last condition is unfortunately not realised in the only data so far available, due to M. Sella, who compared at the same instant, from the summit of Monte Rosa, the brightness of the sun at a height of 40° above the horizon and the brightness of the sky at the zenith. The ratio was found equal to 5 million. Putting this into the formula, and leaving for λ an uncertainty which appears suitable, a value is found for N comprised within

$$30.10^{22} \text{ and } 150.10^{22}.$$

So, in so far as the order of magnitude is concerned, this very interesting theory of Lord Rayleigh is verified, and it is permissible to think that more complete experiments will yield, by this means, a precise determination of N.

38. Direct measurement of the charge of an ion in a gas.— Instead of making an attempt to determine Avogadro's constant or the molecular energy, we may exert ourselves to determine directly the atom of electricity, which, as we have seen, is simply related to them. This is what the physicists of the Cambridge School have succeeded in doing, by determining the charge carried by the ion in gases.

It is not possible to know *à priori* whether, for example, the charge e' of an ion developed in a gas by the passage of X rays bears a simple relation to the charge e which a monovalent ion transports in electrolysis. Naturally, precise measurements of e and e' would decide the question ; but they are not necessary, and to Townsend we owe the establishment, since 1900, by an experimental and theoretical research of extraordinary ingenuity, of the fact that, to about one-hundredth part, the two sorts of ions carry the same charge, the common value e of which it remains to determine otherwise *.

Let us consider ions of the same sign, supposed identical, present in a gas after exposure to X-rays : whatever be their size they will have the same mean kinetic energy as the gas molecules and *will diffuse* in the gas in consequence of this molecular agitation.

Let D be their coefficient of diffusion. On the other hand, let u be the uniform velocity which ions of charge e' assume in this same gas under the influence of an electric field H. N designating always Avogadro's constant, the following equation can be established :

$$Ne' = \frac{RT}{D} \frac{u}{H}.$$

Without reproducing Townsend's reasoning, which makes appeal to some propositions of the kinetic theory not referred to in the course of this Memoir, I will observe that this

---

* *Phil. Trans. of the Royal Soc.* 1900, 129, translated into French in *Ions, Électrons, Corpuscles*, vol. ii. 920 (Gauthier-Villar, publisher).

equation can be very simply obtained by applying to the ions under consideration the formula given by Einstein for the coefficient of diffusion, namely,

$$D = \frac{RT}{N} \frac{1}{6\pi\zeta a}.$$

In reality, applying the law of Stokes to the movement in the electric field, we can write

$$H\epsilon' = 6\pi a \zeta v,$$

and by multiplication, the precise equation of Townsend is obtained.

It is therefore sufficient to know the ratio $\frac{u}{H}$ (or the mobility of the ion), and the coefficient of diffusion D, to determine the product Ne'. Townsend himself has measured this coefficient of diffusion in various gases (air, oxygen, hydrogen, carbon dioxide); using then the measurements of the mobility of ions previously made for these same gases, he has found for Ne' values of which the mean agrees, to less than 1 per cent., with the value 29.10^{13}, fixed by electrolysis for the product Ne. This is a result of primary importance which notably enlarges the ideas, to which electrolysis gives rise, of the existence of an atom of electricity.

But although the first exact demonstration of the invariability of the atomic charge is due to Townsend, Sir J. J. Thomson had already succeeded in showing that the two charges are at least of the same order of magnitude, by attacking directly the measurement in absolute units of the charge e' *. For this he made use of the fact, established by C. T. R. Wilson, that in a moist gas, freed from dust-particles and suddenly supersaturated by the cooling produced by an expansion, the droplets of water form around the ions present in the gas. The method can be summed up as follows :—

---

* *Phil. Mag.* 1898, xlvi. 528; translated into French in *Ions, Électrons, Corpuscles*, vol. ii. 802.

By one of the usual methods the charge E, present in the form of ions per cubic centimetre of the gas, maintained in a constant state of ionisation, is measured, which gives the product ne' of the number of ions present in this volume by the charge e' sought. Then, on suddenly expanding the volume by a known amount, the condensation of a mass of water is brought about, which can be calculated from the known laws of adiabatic expansion. Let m be this mass of water per cubic centimetre of the original gas. If each ion has acted as a germ, this mass is divided between n droplets and, if a is the radius of each droplet, we have

$$m = n \frac{4}{3} \pi a^3.$$

Now the radius a can be obtained from the law of Stokes by measuring the velocity of fall of the cloud under the action of gravity. Since the product ne' is already known, n and in consequence e' can be calculated.

Sir J. J. Thomson has so found for e', in the case of the ions given by X-rays, values between $6·5 \times 10^{-10}$ and $3·4 \times 10^{-10}$, the latter seeming to him the more probable (1903). In the case of the negative ions which are produced by ultraviolet light at the surface of zinc, he found for e' the practically double value $6·8 \times 10^{-10}$. There results from this that the constant of Avogadro ought to be comprised between

$$42.10^{22} \quad \text{and} \quad 85.10^{22},$$

the uncertainty of the mean value being at least 30 per cent. This is the degree of precision of the determination of Van der Waals.

Interesting and instructive as this method is, it contains large sources of error; it is supposed, in particular, that each ion acts as a germ, that each germ only consists of one ion, and that the whole quantity of water calculated has been in reality condensed. The uncertainties are eliminated by an improvement, due to H. A. Wilson [*], who measured the

---

[*] *Phil. Mag.* 1903; translated into French in *Ions, Électrons, Corpuscles*, vol. ii. 1107.

ratio of the velocity of fall of the drops, under the influence of gravity alone, and under the influence of gravity assisted or opposed by a vertical electric field H. We have obviously

$$\frac{v_1}{v_2} = \frac{\frac{4}{3}\pi a^3 g}{\frac{4}{3}\pi a^3 g + He'},$$

a always being given by the law of Stokes :

$$\frac{4}{3}\pi a^3 g = 6\pi \zeta a r_1.$$

H. A. Wilson so found that under the influence of the electric field the charged cloud divided itself into two or even three clouds of different velocities, corresponding to charges which are between themselves as 1, 2, and 3. A drop can therefore absorb many ions, at least unless there are polyvalent ions present in the gas. Further, the value found for e' with the least charged cloud varied notably from one experiment to another, jumping suddenly for example, under conditions apparently identical, from $2·7 \times 10^{-10}$ to $4·4 \times 10^{-10}$. The experiments as a whole indicated, to about 30 per cent. more or less, the value $3·2 \times 10^{-10}$ for e, and in consequence for N the value

$$90.10^{22}.$$

In spite of the ingenuity of the improvements realised by H. A. Wilson, a large uncertainty, therefore, still remains, due perhaps (Rutherford) to the evaporation of the drops during their fall.

However, more recently, some new experiments, made according to the same method by Millikan and Begemann[*], appear to have permitted more accuracy, and give for e the value $4·05 \times 10^{-10}$, and in consequence for N the value

$$72.10^{22},$$

---

[*] *Physical Review*, February 1908, 197.

the uncertainty being possibly only a few per cent. more or less *.

39. Charge of "large ions" present in gases.—The preceding measurements may now be compared with some attempts which have been made to determine the charge, which, according to a process elucidated by Langevin, an ultramicroscopic dust-particle assumes in an ionised gas. Without being able to give here the details of his analysis, it is known that every ion, brought by molecular agitation near to such a dust-particle, is attracted by *the electric image* which it develops in the medium of greater dielectric power, and in consequence sticks to the dust-particle, which remains neutral so long as it receives equal numbers of ions of both signs, but becomes charged when it absorbs an excess of the ions of the one sign. This total charge will therefore be a whole number of electrons, rarely much greater than unity, since when one charge is fixed, ions of the same sign are repelled.

M. Ehrenhaft and M. de Broglie have independently verified these conceptions by most beautiful experiments, no longer following the total displacement of a cloud of particles, but by measuring the individual displacement of these particles †. In their experiments, the air charged

---

* *Note by the Translator.*—Still more recently Prof. Millikan discusses some further results (*Phil. Mag.* Feb. 1910, 209). In the first place, two corrections have to be introduced into the above value for e ($=4 \cdot 05 \times 10^{-10}$), raising it to $4 \cdot 57 \times 10^{-10}$. An important improvement was the use of radium as ionising agent instead of the X-rays, as in Wilson's experiments, which are very variable. Secondly, it has been found possible, by means of an electric field, to hold up *individual* drops carrying multiple charges (2, 3, 4, 5, and 6 respectively) practically stationary in the field of vision for a considerable part of a minute. In this way it was found possible to compare the fall under gravity with the movement, if any, under the electric field *for the same drop*. The mean adopted for all the closely agreeing results is $e = 4 \cdot 65 \times 10^{-10}$, and the uncertainty is reckoned as only about 2 per cent. This makes the value of N

$$62 \cdot 10^{22}.$$

† EHRENHAFT, *Akad. der Wiss. in Wien,* March 1909, and *Physikal. Zeit.* 1909, 308; DE BROGLIE, *Comptes rendus,* May 1909, and *Le Radium,* 1909, 203.

with dust-particles (tobacco-smoke, for example) is drawn through a small transparent box, maintained at constant temperature, where the luminous rays from a powerful source converge. The microscope is placed at right angles to the path of the rays which enables these dust-particles to be seen as very brilliant points animated by a very brisk Brownian movement. When an electric field is made to act at right angles to gravity and to the axis of the microscope, three groups of granules are instantly distinguished: one moving in the direction of the field showing that their charge is positive; another group moving in the inverse direction, which are therefore negatively charged; and, finally, a third group continuing to dance about on the spot and which are therefore neutral.

Precise measurements would be possible if the granules had not, as unfortunately is the case, very varied sizes (and, no doubt, forms). Nevertheless, judging by their brightness, it is possible that they do not depart much from a certain mean diameter deduced, by application of the law of Stokes, from their velocity of fall in a vertical direction *. It only remains to measure the mean speed of the displacement at right angles, in the direction of the electric field, to determine the electric charge, by a second application of the law of Stokes, according to the equation

$$6\pi\zeta av = He.$$

So it is found that the exact value of e cannot differ much from $4\cdot6 \times 10^{-10}$ (Ehrenhaft) or $4\cdot5 \times 10^{-10}$ (de Broglie), which makes N

$$64.10^{22}.$$

In spite of the uncertainty pointed out, I am inclined to consider this method more precise and more easy to perfect than that which depends upon the condensation of drops of water by expansion.

---

* M. de Broglie satisfied himself at the same time, by photographic measurements of the displacements, that Einstein's formula remains, at least approximately, applicable. This confirmation is interesting, in spite of its inferior accuracy compared with that previously obtained with gamboge, because of the great difference of the conditions, especially as regards the viscosity (fifty times feebler in air than in water).

40. Values deduced from radioactive phenomena.—Lastly, an admirable investigation of Rutherford, enlarging still further the idea of the atom of electricity, enables this magnitude to be obtained in many different ways, starting from observations having reference to radioactive bodies [*].

It is known that the α-rays, given off from radioactive bodies, carry charges of positive electricity ; further, that when they strike sulphide of zinc, they develop there small stars of light (*scintillations*) which disappear instantly. These two phenomena have given Rutherford two completely different means of counting the number p of positive projectiles radiated in a second by 1 gram of radium, for the projected particles can reveal their existence *individually*, either by an impulse in an electrometer or by a scintillation. These two methods agree to about 1 or 2 per cent., and give for the required value

$$p = 3 \cdot 4 \times 10^{10}.$$

If, on the other hand, the total charge of positive electricity radiated per second by a given quantity of radium can be measured, a simple division will give the charge e_0 of the α-particle. The measurement is, in fact, difficult, and Rutherford has found for e_0 values comprised between

$$8 \cdot 3 \times 10^{-10} \quad \text{and} \quad 10 \times 10^{-10},$$

that is, about double the atom of electricity. The α-particle is therefore a bivalent ion (in a more precise manner Rutherford has demonstrated that it is a bivalent atom of helium).

The elementary charge e will therefore be one-half of e_0, and taking the mean of the values found

$$4 \cdot 65 \times 10^{-10},$$

the value of N will be

$$62 . 10^{22}.$$

This value is a little less than the value I have found from the study of the Brownian movement.

But Rutherford himself quotes other radioactive facts, which involve equally his fundamental determination of the

---

[*] RUTHERFORD and GEIGER, *Proc. Roy. Soc.* June 1908; translated into French in *Le Radium*, 1908, vi. 257.

loss p of α-particles, radiated by 1 gram of radium, and which lead to numbers practically identical with mine.

One of these relates to an investigation of Boltwood from which it results that the period of transformation of radium can be simply measured and that the transformation is half accomplished in 2000 years. N signifying always Avogadro's constant, and the gram-atom of radium being 226·5 grams, there results from this that the number of atoms of radium which break up per gram during a second, which is probably equal to the number $3·4 \times 10^{10}$ of α-particles radiated during the same time, is also equal to

$$N \frac{1·09 \times 10^{-11}}{226·5},$$

whence it results that the value of N is

$$70·6 \times 10^{22},$$

which is precisely the value I have found.

On the other hand, accepting always for p the number $3·4 \times 10^{10}$ according to Rutherford, it may be supposed that the number of helium atoms produced in one second by 1 gram of radium in radioactive equilibrium is four times $3·4 \times 10^{10}$, since in this radium there are four products, each emitting per second the same number of α-particles, that is to say, atoms of helium. If, therefore, the volume of helium disengaged per second is known, the number of atoms contained in this volume will be known, and in consequence the number of atoms N contained in 1 gram-atom of helium is directly obtained. Now some very careful measurements have been made by Sir James Dewar of the volume of helium disengaged in a day by 1 gram of radium (0·37 cubic mm.) *. As M. Moulin † has observed, this leads to the value of N

$$71.10^{22},$$

which again is practically the value which the study of the

---

* This should have been 0·499 cubic mm. (see Proc. Roy. Soc. 1910, A, lxxxiii. 404). A later result is 0·463. (TRANS.)

† *La valeur la plus probable de la charge atomique* (*Le Radium*, 1909, vi. 164).

Brownian movement has given me. The extraordinary coincidence of the results obtained, by means so profoundly different, is all the more striking because they cannot have exerted any influence, even unconsciously, upon one another; for example, the calculation of M. Moulin was only made after the accomplishment and publication of my researches.

41. Values deduced from the laws of dark radiation.—Lastly, it will not be found less surprising that almost the same numbers have once again been found, starting from relative measurements of the infra-red part of the spectrum of *black* bodies, by the reasoning developed by Lorentz and Planck.

The kinetic theory of metals, such as Sir J. J. Thomson and Drude * have conceived it, has for its fundamental hypothesis the existence in the metals of electric corpuscles, probably identical with those which constitute the cathode-rays, which move in all directions in the metal like the molecules of a gas. Every movement of electricity in a conductor is a movement of these corpuscles as a whole ; but, further, their mean energy of movement increases with the temperature ; being in brisker motion in the hotter regions of the metal, they transmit from place to place this brisker movement by their impacts, and this is what the thermal conductivity of metals consists in. Developing this idea more exactly, Drude supposes that the mean corpuscular energy is equal to the mean molecular energy, and shows that the thermal conductivity should then be proportional to the electric conductivity, which is known as the law of Wiedemann and Franz. Further, the coefficient of proportionality can be calculated *a priori*, and the predicted values agree well with the values which the different metals give.

This remarkable quantitative agreement justifies Drude's hypothesis (and extends, this time into the region of the infinitely small, what we know of the equipartition of energy).

This admitted, Lorentz observes that according to a known

---

* Their memoirs are translated into French in *Ions, Electrons, Corpuscles* (Gauthier-Villars).

law of electromagnetism, these corpuscles which go and come in all directions radiate energy each time their speed changes in direction and magnitude, and according to him, this radiation is precisely and definitively the light which the metal emits at the temperature considered *.

He calculates this radiation, developed in a Fourier series after analysing the emission into rays of different wavelengths, and confines himself after that to waves of which the period is very large in comparison to the mean free path of the corpuscles. For these long waves he calculates similarly the absorbing power of the metal, and obtains (according to Kirchhoff's law) the expression for the dark radiation by dividing the emissive power by the absorbing power †. The result is that, per unit volume, the energy of radiation dA corresponding to wave-lengths comprised between λ and $\lambda + d\lambda$ is

$$\frac{16\pi}{3} w \frac{d\lambda}{\lambda^4},$$

w indicating the corpuscular (or molecular) energy at the temperature considered. This expression may also be written

$$8\pi \frac{RT}{N} \frac{d\lambda}{\lambda^4},$$

and to find N it only remains to measure this energy (which has been done, at least approximately, in the numerous measurements which refer to the distribution of energy in the spectrum of dark bodies). From measurements by Lummer and Pringsheim, Lorentz thus deduces for N the value

$$77.10^{22}.$$

Independently of Lorentz, by a more complicated theory, Max Planck had already arrived at the same formula. The

---

* I may observe that, in this conception, there is really nothing of periodicity in the radiation received in each instant at a point of an isothermal space: *this radiation in equilibrium is a kind of Brownian movement in the ether.*

† His calculations (simplified by Langevin) will be found in the first volume of *Ions, Electrons, Corpuscles*, p. 500.

discussion of the experimental results has led him to a value of N a little different,

$$61.10^{22}.$$

The mean of these two values is **69**.10^{22}, which is very near the preceding values.

There is no doubt that as soon as more certain values of the constant of radiation are available an exact measurement of N will be possible in this direction.

42. Comparison of all the values obtained.—A table will serve to recapitulate usefully the various phenomena which enable N to be calculated, which taken altogether form what may be termed the proof of *molecular reality*.

Phenomena studied.	$N.10^{-22}$
Viscosity of gases taking into account ⎰ the volume of the liquid state	>45
⎱ the dielectric power of the gas	<200
⎰ the exact law of compressibility	60
Brownian Movement. ⎰ Distribution of uniform emulsion	**70·5**
⎱ Mean displacement in a given time	71·5
⎱ Mean rotation in a given time	65
Diffusion of dissolved substances	40 to 90
Mobility of ions in water	60 to 150
Brightness of the blue of the sky	30 to 150
Direct measurement of the atomic charge. ⎰ Droplets condensed on the ions	60 to 90*
⎱ Ions attached to fine dust-particles	64
Emission of α-projectiles. ⎰ Total charge radiated	62
⎱ Period of change of radium	70·5
⎱ Helium produced by radium	71
Energy of the infra-red spectrum	60 to 80

The most probable value always appears to me **70·5** $\times 10^{22}$. The corresponding values of the other molecular magnitudes are given in paragraphs **26** and **27**.

* 62, See Translator's Note, p. 84.

43. Molecular reality.—I think it impossible that a mind, free from all preconception, can reflect upon the extreme diversity of the phenomena which thus converge to the same result, without experiencing a very strong impression, and I think that it will henceforth be difficult to defend by rational arguments a hostile attitude to molecular hypotheses, which, one after another, carry conviction, and to which at least as much confidence will be accorded as to the principles of energetics. As is well understood, there is no need to oppose these two great principles, the one against the other, and the union of Atomistics and Energetics will perpetuate their dua triumph.

Lastly, although with the existence of molecules or atoms the various realities of number, mass, or charge, of which we have been able to fix the magnitude, obtrude themselves forcibly, it is manifest that we ought always to be in a position to express all the visible realities without making any appeal to elements still invisible. But it is very easy to show how this may be done for all the phenomena referred to in the course of this Memoir.

Firstly, so far as concerns each special law, the constant N is simply a completely known numerical factor, figuring in the enunciation of the law. For example, the law of agitation, predicted by Einstein and established in the course of this work, is expressed by the equation

$$\xi^2 = \frac{RT}{7.10^{23}} \frac{1}{3\pi \zeta a},$$

where all the terms are measurable.

But what is perhaps more interesting, and what brings out in some ways what is now tangible in molecular reality, is to compare two laws in which Avogadro's constant enters. The one expresses this constant in terms of certain variables, $a, a', a'', \ldots,$

$$N = f[a, a', a'' \ldots];$$

the other expresses it in terms of other variables, $b, b', b'', \ldots,$

$$N = g[b, b', b'' \ldots].$$

Equating these two expressions we have a relation

$$f[a, a', a'', \ldots] \equiv g[b, b', b'', \ldots],$$

where only evident realities enter, and which expresses a profound connection between two phenomena at first sight completely independent, such as the transmutation of radium and the Brownian movement. For example, if we compare the law of the distribution of the energy A of dark radiation as a function of the wave-length (No. 41) and the law of rarefaction of a uniform emulsion as a function of gravity (No. 14), we perceive that these two laws are not independent and that the one is connected to the other by the equation

$$\frac{1}{dA}\frac{d\lambda}{\lambda^4} = \frac{1}{8\pi}\log\frac{n_0}{n}\frac{1}{(\Delta-\delta)\phi gh},$$

an equation in which all the terms are measurable.

The discovery of such relationships marks the point where the underlying reality of molecules becomes a part of our scientific consciousness.

44. Conclusion.—I think I have given in this Memoir the present state of our knowledge of the Brownian movement and of molecular magnitudes. The personal contributions which I have attempted to bring to this knowledge, both by theory and experiment, will I hope elucidate it, and will show that the observation of emulsions gives a solid experimental basis to molecular theory. The principal results established in the course of this work are in summary :—

The preparation of emulsions with equal spherical granules, of an exactly measured radius, chosen at will ;

The extension of Stokes's law to the domain of microscopic magnitudes ;

The demonstration that the laws of perfect gases apply to uniform emulsions ;

The *exact* determination, from this experimental fact, *of the various molecular magnitudes, and of the charge of the electron* ;

The experimental confirmation, for the rotations as well as for the translations, *of the equipartition of energy*, and of the beautiful theoretical investigations of Einstein ;

Lastly, arising out of this confirmation, a second *exact* determination, agreeing with the first of the various molecular magnitudes.

As we have seen, I was aided in this last part by M. Chaudesaigues, who made with much skill the greater part of the measurements of displacement for gamboge, and who thought of verifying their good agreement with the *law of chance*. On the other hand, I owe to the friendly insistence of M. Dabrowski the repetition, upon a second substance *mastic*, of the first experiments made with gamboge, which has increased their certainty. In these new experiments his able and devoted collaboration has been very useful to me, and I offer him my affectionate thanks. Lastly, I have still to thank M. Dastre, who has been so kind as to put at my disposal the powerful centrifugal machine, which was indispensable for my fractionations.

DOVER PHOENIX EDITIONS

A series of hardcover reprints of major works in mathematics, science and engineering.
All editions are 5⅝ × 8½ unless otherwise noted.

Mathematics

Theory of Approximation, N. I Achieser. Unabridged republication of the 1956 edition. 320pp. 49543-4

The Origins of the Infinitesimal Calculus, Margaret E. Baron. Unabridged republication of the 1969 edition. 320pp. 49544-2

A Treatise on the Calculus of Finite Differences, George Boole. Unabridged republication of the 2nd and last revised edition. 352pp. 49523-X

Space and Time, Emile Borel. Unabridged republication of the 1926 edition. 15 figures. 256pp. 49545-0

An Elementary Treatise on Fourier's Series, William Elwood Byerly. Unabridged republication of the 1893 edition. 304pp. 49546-9

Substance and Function & Einstein's Theory of Relativity, Ernst Cassirer. Unabridged republication of the 1923 double volume. 480pp. 49547-7

A History of Geometrical Methods, Julian Lowell Coolidge. Unabridged republication of the 1940 first edition. 13 figures. 480pp. 49524-8

Linear Groups with an Exposition of Galois Field Theory, Leonard Eugene Dickson. Unabridged republication of the 1901 edition. 336pp. 49548-5

Continuous Groups of Transformations, Luther Pfahler Eisenhart. Unabridged republication of the 1933 first edition. 320pp. 49525-6

Transcendental and Algebraic Numbers, A. O. Gelfond. Unabridged republication of the 1960 edition. 208pp. 49526-4

Lectures on Cauchy's Problem in Linear Partial Differential Equations, Jacques Hadamard. Unabridged reprint of the 1923 edition. 320pp. 49549-3

The Theory of Branching Processes, Theodore E. Harris. Unabridged, corrected republication of the 1963 edition. xiv+230pp. 49508-6

The Continuum, Edward V. Huntington. Unabridged republication of the 1917 edition. 4 figures. 96pp. 49550-7

Lectures on Ordinary Differential Equations, Witold Hurewicz. Unabridged republication of the 1958 edition. xvii+122pp. 49510-8

Mathematical Methods and Theory in Games, Programming, and Economics: Two Volumes Bound as One, Samuel Karlin. Unabridged republication of the 1959 edition. 848pp. 49527-2

Famous Problems of Elementary Geometry, Felix Klein. Unabridged reprint of the 1930 second edition, revised and enlarged. 112pp. 49551-5

Lectures on the Icosahedron, Felix Klein. Unabridged republication of the 2nd revised edition, 1913. 304pp. 49528-0

On Riemann's Theory of Algebraic Functions, Felix Klein. Unabridged republication of the 1893 edition. 43 figures. 96pp. 49552-3

A Treatise on the Theory of Determinants, Thomas Muir. Unabridged republication of the revised 1933 edition. 784pp. 49553-1

A Survey of Minimal Surfaces, Robert Osserman. Corrected and enlarged republication of the work first published in 1969. 224pp. 49514-0

The Variational Theory of Geodesics, M. M. Postnikov. Unabridged republication of the 1967 edition. 208pp. 49529-9

DOVER PHOENIX EDITIONS

An Introduction to the Approximation of Functions, Theodore J. Rivlin. Unabridged republication of the 1969 edition. 160pp. 49554-X
An Essay on the Foundations of Geometry, Bertrand Russell. Unabridged republication of the 1897 edition. 224pp. 49555-8
Elements of Number Theory, I. M. Vinogradov. Unabridged republication of the first edition, 1954. 240pp. 49530-2
Asymptotic Expansions for Ordinary Differential Equations, Wolfgang Wasow. Unabridged republication of the 1976 corrected, slightly enlarged reprint of the original 1965 edition. 384pp. 49518-3

Physics

Semiconductor Statistics, J. S. Blakemore. Unabridged, corrected, and slightly enlarged republication of the 1962 edition. 141 illustrations. xviii+318pp. 49502-7
Wave Propagation in Periodic Structures, L. Brillouin. Unabridged republication of the 1946 edition. 131 illustrations. 272pp. 49556-6
The Conceptual Foundations of the Statistical Approach in Mechanics, Paul and Tatiana Ehrenfest. Unabridged republication of the 1959 edition. 128pp. 49504-3
The Analytical Theory of Heat, Joseph Fourier. Unabridged republication of the 1878 edition. 20 figures. 496pp. 49531-0
States of Matter, David L. Goodstein. Unabridged republication of the 1975 edition. 154 figures. 4 tables. 512pp. 49506-X
The Principles of Mechanics, Heinrich Hertz. Unabridged republication of the 1900 edition. 320pp. 49557-4
Thermodynamics of Small Systems, Terrell L. Hill. Unabridged and corrected republication in one volume of the two-volume edition published in 1963–1964. 32 illustrations. 408pp. 6½ x 9¼. 49509-4
Theoretical Physics, A. S. Kompaneyets. Unabridged republication of the 1961 edition. 56 figures. 592pp. 49532-9
Quantum Mechanics, H. A. Kramers. Unabridged republication of the 1957 edition. 14 figures. 512pp. 49533-7
The Theory of Electrons, H. A. Lorentz. Unabridged reproduction of the 1915 edition. 9 figures. 352pp. 49558-2
The Principles of Physical Optics, Ernst Mach. Unabridged republication of the 1926 edition. 279 figures. 10 portraits. 336pp. 49559-0
The Scientific Papers of James Clerk Maxwell, James Clerk Maxwell. Unabridged republication of the 1890 edition. 197 figures. 39 tables. Total of 1,456pp.
Volume I (640pp.) 49560-4; Volume II (816pp.) 49561-2
Vectors and Tensors in Crystallography, Donald E. Sands. Unabridged and corrected republication of the 1982 edition. xviii+228pp. 49516-7
Principles of Mechanics and Dynamics, Sir William (Lord Kelvin) Thompson and Peter Guthrie Tait. Unabridged republication of the 1912 edition. 168 diagrams. Total of 1,088pp.
Volume I (528pp.) 49562-0; Volume II (560pp.) 49563-9
Treatise on Irreversible and Statistical Thermophysics: An Introduction to Nonclassical Thermodynamics, Wolfgang Yourgrau, Alwyn van der Merwe, and Gough Raw. Unabridged, corrected republication of the 1966 edition. xx+268pp. 49519-1

Engineering

Principles of Aeroelasticity, Raymond L. Bisplinghoff and Holt Ashley. Unabridged, corrected republication of the original 1962 edition. xi+527pp. 49500-0
Statics of Deformable Solids, Raymond L. Bisplinghoff, James W. Mar, and Theodore H. H. Pian. Unabridged and corrected Dover republication of the edition published in 1965. 376 illustrations. xii+322pp. 6½ x 9¼. 49501-9